P9-APK-022

The Arctic Aeromedical Laboratory's Thyroid Function Study: A Radiological Risk and Ethical Analysis

Committee on Evaluation of 1950s Air Force Human Health Testing in Alaska Using Radioactive Iodine [131]

Polar Research Board
Commission on Geosciences, Environment, and Resources
in cooperation with

Board on Health Promotion and Disease Prevention
Institute of Medicine

Board on Radiation Effects Research
Commission on Life Sciences

National Research Council

NATIONAL ACADEMY PRESS
Washington, D.C. 1996

NOTICE: The project that is the subject of this report was approved by the Governing Board of the National Research Council, whose members are drawn from the councils of the National Academy of Sciences, the National Academy of Engineering, and the Institute of Medicine. The members of the committee responsible for the report were chosen for their special competence and with regard for appropriate balance.

This report has been reviewed by a group other than the authors according to procedures approved by a Report Review Committee consisting of members of the National Academy of Sciences, the National Academy of Engineering, and the Institute of Medicine.

Support for this project was provided by the United States Air Force under Contract Number F41624-94-C-2003.

Front: These masks were made by Alaska Natives out of all natural materials that are available in many villages in Alaska. The man was designed by Ruth Rulland and the woman was designed by Rhoda Ahgook.

Copies available from
National Academy Press
2101 Constitution Avenue, N.W.
Box 285
Washington, D.C. 20055
(800) 624-6242
(202) 624-6242 (in the Washington Metro area)
B-704

Library of Congress Catalog Card No. 95-72623
International Standard Book No. 0-309-05428-1

Printed in the United States of America

COMMITTEE ON EVALUATION OF 1950s AIR FORCE HUMAN HEALTH TESTING IN ALASKA USING RADIOACTIVE IODINE [131]

CHESTER M. PIERCE, *Chair,* Harvard Medical School, Boston, Massachusetts
DAVID BAINES, St. Maries Family Medicine Clinic, St. Maries, Idaho
INDER CHOPRA, University of California at Los Angeles, School of Medicine, Los Angeles, California
NANCY M. P. KING, University of North Carolina School of Medicine, Chapel Hill, North Carolina
KENNETH L. MOSSMAN, Arizona State University, Tempe, Arizona

Staff

CHRIS ELFRING, Polar Research Board, Study Director *(after July 1995)*
LOREN W. SETLOW, Polar Research Board, Study Director *(through July 1995)*
TONI GREENLEAF, Polar Research Board, Senior Project Assistant
MICHAEL STOTO, Board on Health Promotion and Disease Prevention
JOHN ZIMBRICK, Board on Radiation Effects Research

COMMISSION ON GEOSCIENCES, ENVIRONMENT, AND RESOURCES

M. GORDON WOLMAN, *Chair*, The Johns Hopkins University, Baltimore, Maryland
PATRICK R. ATKINS, Aluminum Company of America, Pittsburgh, Pennsylvania
JAMES P. BRUCE, Canadian Climate Program Board, Ottawa
WILLIAM L. FISHER, University of Texas at Austin
GEORGE M. HORNBERGER, University of Virginia, Charlottsville
DEBRA KNOPMAN, Progressive Foundation, Washington, D.C.
PERRY L. MCCARTY, Stanford University, California
JUDITH E. MCDOWELL, Woods Hole Oceanographic Institution, Massachusetts
S. GEORGE PHILANDER, Princeton University, New Jersey
RAYMOND A. PRICE, Queen's University at Kingston, Ontario
THOMAS C. SCHELLING, University of Maryland, College Park
ELLEN SILBERGELD, University of Maryland Medical School, Baltimore, Maryland
STEVEN M. STANLEY, The Johns Hopkins University, Baltimore, Maryland
VICTORIA J. TSCHINKEL, Landers and Parsons, Tallahassee, Florida

NRC Staff

STEPHEN RATTIEN, Executive Director
STEPHEN D. PARKER, Associate Executive Director
MORGAN GOPNIK, Assistant Executive Director
GREGORY SYMMES, Reports Officer
JAMES E. MALLORY, Administrative Officer
SANDRA S. FITZPATRICK, Administrative Associate
SUSAN SHERWIN, Project Assistant

The National Academy of Sciences is a private, nonprofit, self-perpetuating society of distinguished scholars engaged in scientific and engineering research, dedicated to the furtherance of science and technology and to their use for the general welfare. Upon the authority of the charter granted to it by the Congress in 1863, the Academy has a mandate that requires it to advise the federal government on scientific and technical matters. Dr. Bruce M. Alberts is president of the National Academy of Sciences.

The National Academy of Engineering was established in 1964, under the charter of the National Academy of Sciences, as a parallel organization of outstanding engineers. It is autonomous in its administration and in the selection of its members, sharing with the National Academy of Sciences the responsibility for advising the federal government. The National Academy of Engineering also sponsors engineering programs aimed at meeting national needs, encourages education and research, and recognizes the superior achievements of engineers. Dr. Harold Liebowitz is president of the National Academy of Engineering.

The Institute of Medicine was established in 1970 by the National Academy of Sciences to secure the services of eminent members of appropriate professions in the examination of policy matters pertaining to the health of the public. The Institute acts under the responsibility given to the National Academy of Sciences by its congressional charter to be an advisor to the federal government and, upon its own initiative, to identify issues of medical care, research, and education. Dr. Kenneth I. Shine is president of the Institute of Medicine.

The National Research Council was organized by the National Academy of Sciences in 1916 to associate the broad community of science and technology with the Academy's purposes of furthering knowledge and advising the federal government. Functioning in accordance with general policies determined by the Academy, the Council has become the principal operating agency of both the National Academy of Sciences and the National Academy of Engineering in providing services to the government, the public, and the scientific and engineering communities. The Council is administered jointly by both Academies and the Institute of Medicine. Dr. Bruce M. Alberts and Dr. Harold Liebowitz are chairman and vice chairman, respectively, of the National Research Council.

Erratum

The Arctic Aeromedical Laboratory's Thyroid Function Study began in 1955, not 1956, as inadvertently stated on pages vii, 1, and 9.

Preface

With the end of the Cold War has come more freedom to step back and examine its legacy. In May 1993, experts and public officials met at the Arctic Contamination Conference in Anchorage, Alaska, to discuss problems relating to the post-World War II era of human occupation of the Arctic. The conference was sponsored by 14 federal agencies that participate in the Interagency Arctic Research Policy Committee, and it focused attention on radiation and chemical experimentation and contamination of the Arctic.

One issue raised at the Anchorage meeting was a 1956-1957 study conducted by the U.S. Air Force's former Arctic Aeromedical Laboratory (AAL) about the role of the thyroid gland in acclimatization of humans to cold. The study used $Iodine^{131}$ (I^{131}), a radioactive medical tracer, to measure thyroid activity in 121 people—102 Alaska Natives and 19 military personnel. When the research came to light at the conference, serious questions were raised: How were the research subjects selected? Did the subjects fully understand the purpose of the research? Were they informed of the risks? To help resolve the controversy, Congress, under the leadership of Senator Frank Murkowski (R-Alaska), asked the National Research Council (NRC) and Institute of Medicine (IOM) to review the AAL thyroid function study (Public Law 103-160). The National Research Council and Institute of Medicine (NRC/IOM) appointed the Committee for Evaluation of Air Force 1950s Human Health Testing in Alaska Using Radioactive $Iodine^{131}$ to fulfill that request. The Committee was charged to investigate whether the study was conducted in accordance with generally accepted guidelines in the 1950s for use of human participants in medical experimentation and whether the I^{131} doses used were administered in accordance with radiation exposure standards of the 1950s, as well as how the dosages would compare to modern standards. The Committee was also asked to examine whether the AAL thyroid function study had followed guidelines regarding informing participants about possible risks and whether subsequent studies of the participants should have been conducted to determine whether any suffered long-term ill effects.

Given that the events at issue occurred some 40 years ago, the Committee's charge was not an easy one. The Committee first met by teleconference in June 1994 and began an intensive effort to gather records and information about the research and to locate the subjects tested. One Committee member (a physician) and an NRC staff member then traveled to two rural Alaska villages in July 1994 to interview Native subjects. Immediately following those field interviews, the full Committee met in Fairbanks, Alaska, to hear from a wide range of people at a public hearing. Speakers included Native study participants, a military participant, doctors who worked at the Air Force laboratory in the 1950s, the physician that led the study (by phone from Norway), a medical historian, and representatives of state, local, and tribal government agencies. This report is the Committee's careful analysis of this information and its best judgment about the difficult issues that this research brings to light.

The Committee would like to thank all those who participated in the hearing and other information-gathering activities—this personal input was critical. The Committee did its best to sort through the sometimes wide-ranging testimony and to focus on the AAL thyroid function study. It was clear from the public hearing, however, that many Alaskans are concerned with issues broader than this one study—questions about other experiments on Alaska Natives. Many at the public hearing expressed great frustration, and even rage, at the lack of information and explanation of what was done to them during this and other past research. Yet despite this anger, they were exceedingly generous in their willingness to help the Committee in its investigation.

While the Committee's review of the AAL thyroid research was ongoing, information about other medical studies of the 1950s was coming to light. One important effort was undertaken by the President's Advisory Committee on Human Radiation Experiments (ACHRE). ACHRE was established in 1994 to provide advice on ethical and scientific issues related to government-sponsored human radiation experiments, that is, experiments designed to understand the impacts of intentional exposure to ionizing radiation (excluding common and routine clinical practices) and experiments involving intentional environmental releases of radiation. ACHRE has raised significant questions about the conduct of such experiments (Advisory Committee on Human Radiation Experiments, 1995), and, although the use of a radioactive medical tracer presents far less risk and involves different objectives, the lessons from the ACHRE report have some bearing here.

The AAL thyroid research must be seen in the context of the era when it was conducted. The methodology was not unusual and the researchers interacted with the people who served as subjects in the manner and tone that was typical of the times. The diagnostic use of I^{131} was routine in medical practice across America in the 1950s. Research has not shown any link between small doses of I^{131} and thyroid cancer. Given these facts, why should a single study using diagnostic amounts of the tracer cause so much concern? In a June 1994 letter to the Committee, Senator Murkowski explained: "Unfortunately, the $Iodine^{131}$ tests almost 40 years ago—even if they were safe, well intentioned, and conducted in accordance with the standards of the day—have contributed to an atmosphere of conflict and mistrust between the indigenous peoples of the Arctic and the community of scientists and researchers who work in the Arctic. That is troubling and unfortunate, because science is critically important to Alaska and its future." He hoped an honest appraisal of the AAL thyroid function study would improve the relationship between the scientific and indigenous communities.

The Committee bears full responsibility for the content and opinions presented in this document. However, we would like to acknowledge the many people who provided assistance. The Committee would like to thank the NRC staff for their support, including their diligent search for archival records and willingness to pursue cold trails for relevant information. The Committee also must thank the U.S. Air Force's Office of the Surgeon General, the Mayor of the North Slope Borough and his staff, the Tanana Chiefs Conference in Fairbanks, and the Indian Health Service for their assistance in searching for and locating medical study subjects, and the U.S. Air Force for its financial support of the project. The Committee also received valuable assistance from the Oak Ridge Associated Universities Institute for Science and Education in calculating thyroid gland doses from administered I^{131} and from many others who spoke to us, reviewed our drafts, and provided other input. In addition, the Committee wishes to thank ACHRE for helping us review historical documents related to the conduct of medical research by the U.S. military in the 1950s. And, finally, I would like to thank personally the members of the Committee for their devoted participation in this activity.

Although it is difficult and relatively unsatisfying to comment after the fact on what should have been done at a point some 40 years in the past, the Committee felt that this investigation into past actions is important. Lessons from the past can lead to better research today, especially when it comes to cross-cultural studies. In addition, understanding the past can help set aright the continuing and long-term effects of a perceived betrayal of trust between the Alaska Native peoples and the incoming, dominant culture—represented in this case by the scientific community. This study and the public dialogue that occurred in the course of the study are a positive step in making complete information available and meeting modern standards of accountability.

Chester Pierce, *Chair*
Committee on Evaluation of 1950s Air Force
Human Health Testing in Alaska Using Radioactive Iodine^{131}

Contents

6 APPENDICES

Executive Summary

During the 1950s, the potential for nuclear war between the United States and the Soviet Union was a major concern, and Alaska was seen as the likely battleground because of its proximity to the Soviet mainland. In the pursuit of military preparedness, many activities ensued—troop deployment, facilities development, weapons testing, and medical research—to ensure that troops would be ready to operate in the challenging environment.

In 1956 and 1957, the U.S. Air Force's former Arctic Aeromedical Laboratory (AAL), a research facility charged to investigate ways to prepare fighting forces to deal with the harsh climate, conducted a study of the role of the thyroid gland in human acclimatization to cold. To measure thyroid function under various conditions, the researchers administered a medical tracer, the radioisotope $Iodine^{131}$ (I^{131}), to 121 people—102 Alaska Native subjects and 19 white military personnel—and measured its uptake and excretion. Based on the study results, the AAL researchers determined that the thyroid did not play a significant role in human acclimatization to extreme cold.

The AAL thyroid function study came to light again in 1993, when it was discussed at a conference on the Cold War legacy in the Arctic. The study proved to be quite controversial, and questions were raised about whether it had posed risks to the people involved and whether the research had been conducted within the bounds of accepted guidelines for research using human participants. At the request of Congress (P.L. 103-160), the National Research Council agreed to undertake an analysis and appointed a committee, the Committee on Evaluation of 1950s Air Force Human Health Testing in Alaska Using Radioactive $Iodine^{131}$, to conduct this study. This report reviews the purpose and methodology of the AAL thyroid function study, assesses the medical risks from using diagnostic level doses of the radioactive tracer, examines standards for the use of these tracers then and now, examines the ethics of human subjects research from both a 1950s and a 1990s perspective, and presents the Committee's conclusions and recommendations.

DESCRIPTION OF THE AAL THYROID FUNCTION STUDY

The AAL, which was established in 1951 in Fairbanks, Alaska, and ceased operations in 1967, was responsible for a broad range of studies to improve our understanding of human adaptation to the Arctic environment. Studies dealt with nutrition, physiology, and other elements of cold weather acclimatization to help military personnel fight in the harsh climate and be prepared to survive in cold, barren territory in case of emergencies. The AAL thyroid function study ("Thyroid Activity in Men Exposed to Cold," Rodahl and Bäng, 1957) was one study from the laboratory's portfolio. The study's purpose was to determine whether the thyroid played a role in human adaptation to cold, and the research subjects included Alaska Native men, women, and children in a number of villages in northern and central Alaska and Air Force and Army servicemen. In the research, capsules of radioisotope I^{131} were administered and the radioiodide uptake in the subject's thyroid, blood, urine, and saliva was measured. The study was conducted in different seasons for Alaska Native subjects and before and after exposures to cold stress for the military personnel. Overall, the study included six different tests and lasted from August 1955 to February 1957. A total of 200 doses were distributed to 121 research subjects, with most doses being 50 microcuries (as was standard for I^{131} tracer studies at the time). To locate Alaska Native subjects, the researchers approached village elders or leaders and explained the study, and these leaders then solicited volunteers. Military volunteers were obtained via requests from their commanding officers.

In the end, the AAL thyroid function study concluded that there was no significant difference in thyroid uptake or urinary elimination of radioactive tracers between the Alaska Natives and the white military personnel tested. Nor was there any indication of increased thyroid stimulation as a result of exposure to cold. There was no racial difference in the uptake or excretion of the tracer nor any consistent seasonal variation. The thyroid, the researchers concluded at the time, did not play a significant role in human acclimatization to the Arctic environment. (In the 1960s, when more sophisticated techniques existed, research did determine some relation between thyroid hormone production and metabolic clearance in the body in response to cold and lengthening daylight.)

HEALTH EFFECTS OF I^{131} ADMINISTRATION IN HUMANS

I^{131} was the only radioactive tracer readily available for use in the 1950s, when the AAL thyroid function study was conducted. It is rarely used in modern nuclear medicine because of the relatively high radiation dose received by the thyroid and because more suitable radioisotopes are now available. The AAL use of 50 microcurie doses was standard and licensed by the U.S. Atomic Energy Commission; the AAL principal investigator received standard training in the use of the technique.

The probability of radiation-induced thyroid cancer, or risk, is the product of the radiation dose to the thyroid and the absolute risk coefficient (excess number of cancers per million persons per rad). Using the best available data and methodology, the Committee calculated the thyroid cancer risk to the Alaska Natives and white military personnel who participated in the AAL study. The lifetime estimates of the probability of thyroid cancer as a result of I^{131} administration during the AAL thyroid function study are listed in Chapter 2, Table 2.3. As calculated in Chapter 2, the weighted average risk among the various populations who participated is about 1 in 3,000, a risk six times lower than the background thyroid cancer risk. Because of gender and age differences and the fact that the subjects received a wide range of doses, the risk estimates vary among subjects. The lowest doses and risks were seen in Wainwright, Point Lay, Fort Yukon, and Point Hope Alaska Natives, the majority of whom received single doses. The highest doses and risks appear in Anaktuvuk Pass Eskimos and Arctic Village Indians, who received multiple administrations of I^{131}. For instance, the greatest risk (albeit small) is to Anaktuvuk Pass and Arctic Village females who received multiple doses (calculated to be 1 in 800 and 1 in 700, respectively).

To put the risks into perspective, it is important to consider the natural incidence of thyroid cancer in the population and the lifetime risk of thyroid cancer in the absence of radiation exposure. Thyroid cancer is a rare form of cancer—it is estimated there will be 14,000 new cases in the United States in 1995 (compared, for instance, to 183,000 new cases of breast cancer and 170,000 new cases of lung cancer). This is about 5 cases per 100,000 population. Assuming a 40-year period of risk, the total lifetime background thyroid cancer risk is about 200 per 100,000, or 1 in 500. The weighted average excess risk due to the I^{131} study among the various populations, calculated to be about 1 in 3,000, is six times lower than the background thyroid cancer risk. Thus, participation in the AAL study added an extremely small amount to the background thyroid cancer risk. Even those subjects who received multiple doses, and thus the greatest risk of all those who participated, face risks lower than the background risk. Given that the study had only 121 participants, radiation-induced thyroid cancers caused by the AAL study would not be expected in either the Alaska Natives or military personnel who served as research subjects. In fact, thyroid cancer has not been observed in any member of the study population. (In addition, some health benefits may have been coincidentally provided because the researchers identified endemic goiter problems in Arctic Village and Anaktuvuk Pass.)

At the time the AAL study was conducted in the mid-1950s, there were no formal guidelines concerning radiation exposure of research subjects. The Atomic Energy Commission approved the study based on radiological considerations and I^{131} was the only material readily available to conduct the study. In addition, the prevailing view at the time was that radiation effects required that the dose exceed a threshold level, leading researchers to believe that radiation doses below the threshold were safe. Thus, from a radiation exposure perspective, the study was scientifically reasonable by the standards of the time. However, the same activity would be unlikely to be approved under current standards because the doses given exceed the current recommended dosage limits. Current standards also would preclude use of minors, lactating women, and potentially pregnant women. The use of subjects with thyroid enlargement would also be a questionable practice.

THE ETHICS OF HUMAN SUBJECTS RESEARCH

The ethical guidelines for the conduct of human subjects research have evolved over time. Determining whether the AAL thyroid function study was conducted according to generally accepted guidelines of the 1950s requires an understanding of just what those guidelines were and how they were applied.

Before World War II, some attention was paid to the ethical issues raised by medical research with human subjects; it focused more on controlling research risks than on enabling autonomous choice by research subjects. But standards were evolving, and this evolution was brought to public consciousness by the Nuremberg Trials and development of the Nuremberg Code, a watershed in bioethics. The Code, a 10-point list of principles delimiting morally and legally permissible human experimentation, was issued in August 1947 and was intended to be an expression of universal moral principles governing research with human subjects. Among other things, it requires that subjects have decision-making capacity; that they be able to consent freely, without intervention of force or other forms of coercion; and that they be given information about the nature and purpose of the experiment and about all reasonably anticipated risks. The Code was well promulgated and widely discussed in the late 1940s and early 1950s.

But in the postwar period, dissemination of and implementation of the Nuremberg Code was, to say the least, uneven. In 1953, the Department of Defense formally adopted the Code in guidelines addressing the use of human subjects for research related to atomic, biological, and chemical warfare but the document was classified Top Secret because of government sensitivity on these military issues. (This is the earliest known instance in which a federal agency that sponsored human experiments adopted the Nuremberg Code.) The Atomic Energy Commission developed subject consent requirements for the use of radioisotopes but did not systematically promulgate or enforce them. Many medical researchers apparently believed that because the Nazi experiments were so egregiously flawed, both ethically and scientifically, the Nuremberg Code was intended to apply only to ill-intentioned research. Nonetheless, ethical guidance for physicians and researchers continued to evolve. For instance, in 1953 the National Institutes of Health implemented a rigorous policy requiring informed consent and peer review of risk-bearing research in its clinical center. Thus, in the early to mid-1950s the principles governing research with human subjects were firmly in place, but their implementation in practice was incomplete and even confused. Clear moves toward systematic reform of research practice did not come until the 1960s and 1970s.

Despite the unevenness of the application of ethical standards for the conduct of human subjects research in the 1950s, this Committee concludes that the standards outlined in the Nuremberg Code did apply to human subjects research at the time, including research conducted under military auspices and the AAL thyroid function study. In comparison with some other Cold War research, such as some of the experiments examined by the Advisory Committee for Human Subjects Research (ACHRE), which include examples of actual deception of patient-subjects, it might seem overly scrupulous to be so concerned with the AAL application of I^{131}. Nevertheless, there are three reasons for careful examination of the study according to the terms of the Code: (1) the subjects were normal healthy volunteers and there was no indication that the researchers expected the study to improve the subjects' health; (2) radioisotopes are potentially harmful substances, so the research was not without theoretical risk in spite of the fact that the investigators thought there was no risk; and (3) the majority of the subjects were Alaska Natives, whose cultural and language differences affected the consent process.

Based on its analysis, the Committee concludes that information on the nature of the I^{131} tracer was not fully disclosed to the research subjects, and that therefore the military and Alaska Native subjects were not completely informed about the nature and risks of the experiments. In the case of Alaska Native subjects, the researchers accepted as volunteers anyone brought to them without inquiring as to what the subjects had been told, and they relied on elders or other intermediaries without medical or scientific training to obtain volunteers and explain the research. Minor children were used without adequate parental consent. Few of the Alaska Native subjects understood that they were participating in research; instead, most thought they were receiving medical treatment. Neither the Alaska Natives nor the military personnel were informed about the radioactive tracer. Because of these deficiencies, the Committee believes that the experiments were conducted without informed consent, even according to the standards of the time. The AAL experimental design and consent process also clearly falls short of modern standards.

CONCLUSIONS AND RECOMMENDATIONS

For every research project involving human subjects, two basic inquiries are necessary. One inquiry must examine the necessity of the research, the expected results, the risk-benefit balance, and minimization of risk. The other must examine the fairness of subject selection, adequacy of information given to prospective subjects, and the voluntariness of the subjects' consent to participation. In general terms, the first inquiry addresses the research's potential for *harming* subjects and the second addresses the research's potential for *wronging* them. The first concept is based on benefits; the second is concerned with autonomy and justice. The two concepts are interdependent, but it is nonetheless possible to commit harm without wrong, and wrong without harm.

After examining the records, analyzing the health risks, and talking with research subjects as well as researchers, the Committee concludes that the probability of physical harm to the AAL study subjects is negligible, and thus that the subjects were not harmed. From an ethical perspective, the Committee concludes that aspects of the AAL study, especially the informed consent process, were flawed even by 1950s standards and thus the Alaska Natives who participated and, to a lesser extent, the military subjects were wronged. Although wrong was done, it is vital to emphasize that it is inappropriate to place blame. The researchers were conscientious scientists who held a genuine belief, justified at the time, that their research was both harmless and important. The research design was approved by their superiors. The lack of emphasis on autonomy and informed consent, and the lack of cultural sensitivity, were standard errors of the time. It is the Committee's hope that acknowledgement of these wrongs will reduce the likelihood of similar wrongs in the future, and that open discussion of them will enhance the level of trust between the people involved and the government.

The U.S. Air Force, U.S. health organizations, and Congress should take steps to redress the wrong done by the researchers' failure to obtain informed consent during the AAL thyroid function study. The first step is providing information on the true magnitude of the risks and possible consequences of the research to surviving subjects, their families, and their villages. In this spirit, the Committee recommends the following:

(1) The government and the Air Force should acknowledge responsibility for wrongs done in the course of the AAL thyroid function study in the hopes of ensuring that similar problems do not occur in the future, and they should address the wrongs by undertaking the following actions:

(a) The Air Force should endeavor to contact all living subjects or their immediate families and provide records to them of their AAL research participation in the I^{131} experiments. The Air Force should also continue to search for records of the AAL that would identify the six U.S. Army subjects and six Point Hope subjects who were not named in the Air Force report of the study, and to locate the Air Force and Army subjects named in the study.

(b) In the process of contacting subjects and subjects' families, the Air Force should disseminate the Committee's report and other available information on human medical experimentation conducted by the AAL in the period 1948 - 1967 to appropriate health care providers, tribal governments, and other key figures in the relevant Alaska Native villages.

(2) U.S. government and Alaska state health organizations, under U.S. government auspices, could complement the efforts of the Air Force by conducting related public health education programs facilitated by Native experts focused on conveying information about patients' rights in any therapeutic or research situation, and medical information about exposure to radiation. Such a process will enable Native experts, clinics, and physicians to provide accurate information to their communities and ease fears.

(3) If Congress considers legislation to redress any wrongs or harms done to human subjects of government radiation research where informed consent was not obtained, the Committee believes Congress should consider including the subjects of the AAL thyroid function study.

(4) Although medical follow-up based on the calculated risk values is not warranted, the U.S. Air Force should provide medical follow-up to those participants who were under age 20 at the time of the AAL study since those participants will be at risk for the longest period of time. Such follow-up would provide assurance that these participants suffered no long-term physical ill effects.

The Committee recognizes that its basic conclusion—that the subjects of the AAL thyroid function study were wronged but not harmed—may prove controversial. Some will claim that the Committee's calculations are incorrect and that the risk is higher. Others will believe that the Committee failed to go far enough in suggesting ways to right the wrongs. Some will say that the Committee failed to understand the climate of the times—the intensity of the Cold War pressures and national security concerns and the fact that many researchers truly did not believe that the Nuremberg Code applied to benign human subjects research. They may claim that the Committee was swayed by the clarity that only hindsight brings.

The Committee believes that these various perspectives arise from concern for the people involved, both the researchers and their superiors and the research subjects. It recognizes that some subjectivity is inherent in this type of analysis and that honest differences of opinion can occur. Still, the Committee is convinced that its position is defensible, sensible, and ethical. The risk analysis in this report is based on the best epidemiology and dosimetry available. It is, if anything, conservative, and the real risk may actually be smaller than expressed. The Committee's position acknowledges the flaws of the AAL thyroid function study within the context of history, while not placing blame on those who conducted the activity using what they perceived to be harmless methods in pursuit of justifiable goals.

1

Introduction

The United States first established a significant military presence in Alaska in 1942 after the Japanese bombed and occupied islands in the Aleutians. In the late 1940s and early 1950s, as Cold War pressures escalated, the United States feared a nuclear confrontation with the Soviet Union and the U.S. military presence in the Arctic grew. Long-range bombers, fighter squadrons, and early-warning radar stations were set up in remote corners of Alaska to prepare for the possibility of a confrontation between the two superpowers.

With the Cold War looming, military planners sought to know more about how to keep fighting forces fit and capable in extreme cold and how to survive in emergencies in the harsh climate. In 1951, the Air Force established the Arctic Aeromedical Laboratory (AAL) in Fairbanks, Alaska, to conduct research on Arctic acclimatization. The U.S. Air Force[1] sought information on medical problems related to cold for three main reasons: (1) aircraft were attaining higher altitudes, exposing military personnel to colder temperatures; (2) transpolar flight was becoming increasingly important, and flight crews had to know how to survive in cold and barren territory in case of emergencies; and (3) flight maintenance crews based in polar regions had to work efficiently and accurately in cold conditions. The AAL's mission was "... to conduct research and development into the human factors problems incident to U.S. Air Force activities in Arctic and Sub-Arctic regions for the purpose of increasing operational capability of the U.S. Air Force in th(o)se regions." The AAL[2] tackled these problems with two distinct, parallel approaches: first, fundamental medical research, and second, the development and field-testing of arctic survival equipment and methods of evacuation and treatment of casualties. The AAL's mission came to include the study of natural food sources;

[1]Information on the history of the AAL comes from a report of the U.S. Air Force (undated).

[2]The AAL was located at Ladd Field in Fairbanks, and was under the operational control of the Alaskan Air Command and its successors, but the laboratory's commanding officer also reported directly to the Surgeon General of the U.S. Air Force. The AAL ceased operations in 1967, and its functions were transferred to Brooks Air Force Base, San Antonio, Texas.

dietary requirements; survival aids; acclimatization to cold; and the adequacy of survival equipment, rations, and clothing.

One of the AAL's research projects, conducted in 1956 and 1957, investigated the role of the thyroid in acclimatization to cold (see Box 1.1). Based on previous research, including animal studies, researchers at the AAL had hypothesized that the thyroid might play a role in cold weather survival and designed a study to measure the effects of cold on thyroid activity by comparing Alaska Native subjects and white military personnel. To measure thyroid function, researchers used a radioactive medical tracer, the radioisotope iodine 131 (I^{131}). This tracer was developed in the 1940s to track absorption of iodine by the thyroid, its expression in the thyroid hormones in blood and tissues, and excretion by the body. It was commonly used in the 1950s (and, in fact, is still used today for certain purposes). In all, 121 people—102 Alaska Native subjects and 19 military subjects—were administered 200 doses of I^{131}. Based on the study, researchers determined that the thyroid did not play a significant role in human acclimatization to extreme cold; they then published their results in the scientific literature and moved on to other projects.

It was not until a 1993 conference on the Cold War legacy in the Arctic that the AAL thyroid function study garnered further attention. When the study came to light, questions were raised about the appropriateness of the activity—whether it posed risks to the people involved and whether the research had been conducted within the bounds of accepted guidelines for research using human participants. In particular, there was concern over the relatively large number of Alaska Natives used as subjects and whether the research subjects understood the nature of the study. As a result, Congress asked the National Research Council (NRC) and Institute of Medicine (IOM) to conduct an evaluation of the AAL's thyroid function research project.

THE COMMITTEE'S CHARGE

The NRC and IOM appointed the Committee on Evaluation of 1950s Air Force Human Health Testing in Alaska Using Radioactive Iodine 131 to evaluate the AAL thyroid function study. The committee was charged to address four key issues[3]:

(1) Whether the series of medical studies was conducted in accordance with generally accepted guidelines in the 1950s for use of human participants in medical experimentation.

(2) Whether the I^{131} doses used in the studies were administered in accordance with radiation exposure standards generally accepted as of 1957 and how those dosages compare to the radiation exposure standards accepted as of 1993.

(3) Whether the studies had and followed guidelines regarding notification of participants about any possible risks.

---

[3]This study was authorized by Congress in Public Law 103-160, the National Defense Appropriations Act for Fiscal Year 1994 (approved November 30, 1993). The tasks listed were assigned by Congress, although the language has been edited for clarity.

Box 1.1
THE THYROID GLAND

The thyroid is a shield-shaped gland located at the base of the neck that produces secretions essential to regulate metabolism, the basic chemical changes in body cells by which energy is provided for life processes and activities. It is also important in the growth of the human fetus and infant. The thyroid gland is one of several endocrine glands characterized by their ability to synthesize and secrete hormones. Located at the four corners of the thyroid gland are four parathyroid glands the size of pearls; the parathyroid glands produce parathyroid hormone, which works with another hormone called calcitonin, also made by the thyroid, and vitamin D to control the level of calcium in the blood.

The prime hormones produced by the thyroid are known as thyroxine (T_4) and 3,5,3-triiodothyronine (T_3). Methods to measure the hormones T_3 and T_4 in the blood were not available in the 1950s. (If they had been, they would have shown that there is a difference in thyroid hormone use in the body due to cold.) It was known, however, that iodine constitutes an important component of thyroid hormones and that the thyroid gland was capable of concentrating iodine from the blood and converting iodine into thyroid hormone. The thyroid hormone is ultimately released into circulating blood, where it affects cell metabolism.)

(4) Whether subsequent studies of the participants should have been conducted to determine whether any participants suffered long-term ill effects from the administration of I^{131} and, in the case of ill effects, whether medical care for such effects was needed.

To conduct this study, the Committee sought to examine as much written documentation as possible, including AAL historical records, results, procedures of experiments, and case files of participants and investigators. The Committee also examined historical and current literature on the use of radioisotopes for diagnosis and therapy of thyroid dysfunction and federal and professional guidelines for the conduct of human biomedical experimentation. In addition, extensive oral and written testimony was considered from those research participants and investigators who could be located.

With this base of information, the Committee sought to assess the conduct of the experiments, the radiological dosage levels, and the risks to the participants. It sought to determine whether, and how, the participants were informed of the nature of the research and the usage of the radioactive isotope. Finally, the Committee considered the evolution of the concept of "informed consent" and how that concept would have been applied in the 1950s as opposed to today's standards. The Committee recognizes the uncertainties inherent in assessing events that took place 40 years ago, but has done its best to present a complete analysis.

This chapter reviews the purpose and methodology of the AAL thyroid function study and introduces the Committee's methods of gathering and analyzing information. Chapter 2 assesses the medical risks from using the diagnostic level doses of the radioactive tracer and examines standards for the use of these tracers then and now. Chapter 3 considers the ethics of human subjects research, from both a 1950s and a 1990s perspective. The Committee's conclusions and recommendations are presented in Chapter 4. A selection of supporting materials is included in the appendices.

THE AAL AND THE THYROID FUNCTION STUDY

To understand the nature and purpose of the AAL thyroid function study, it is important to see it in context. The AAL was conducting a broad range of studies during the 1950s to improve our understanding of human adaptation to the Arctic environment. The researchers included both nonmilitary and U.S. Air Force doctors and scientists, and research subjects included animals, military personnel, and Alaska Natives. One laboratory project, entitled "Human Acclimatization and Adaptation to Arctic Cold," alone included 35 separate studies (U.S. Air Force, 1957b). The AAL thyroid function study, the subject of this report, was one study in this portfolio.

From 1950 to 1957, the human acclimatization research was run by Dr. Kaare Rodahl, a Norwegian physician and Fellow of the University of Oslo, who was recruited by the U.S. Air Force to direct the AAL's Department of Physiology, and later all of its research, because of his extensive (and relatively rare, at the time) expertise in arctic medicine. Dr. Rodahl oversaw a variety of studies on the nutrition, physiology, and living habits of Alaska Natives from villages throughout Alaska (Rodahl, 1952, 1954; Rodahl and Rennie, 1957; Meehan, 1955; Drury et al., 1956). None of the studies used radioactive substances, and most were noninvasive. They consisted of diet and lifestyle histories and the performance of physical examinations (including chest x-rays in some instances); measurements included sampling of blood and urine and analysis of diet samples.

Despite the importance of its mission, the resources available to the AAL were relatively limited. Because of the AAL's remote location, obtaining specialized supplies for scientific experiments was always a problem. The laboratory's operations were also limited by transportation difficulties, the harsh climate, and the dispersed population of interior and northern Alaska.

Researchers at the AAL suspected that the thyroid gland played a role in cold weather survival. Using animal studies, researchers determined that animals exposed to cold for prolonged periods show changes in thyroid structure and function (Leblond and Gross, 1943; Leblond et al., 1944). They concluded that hyperactivity of the thyroid reflected increased metabolism in cold (Therien, 1949), and suggested that the thyroid was involved in human acclimatization to cold (Brown and Hatcher, 1953).

Thus, from 1955 to 1957, the AAL conducted a study designed to explore whether the thyroid played a role in human adaptation to cold—"Thyroid Activity in Men Exposed to Cold" (Rodahl and Bäng, 1957). The study used the radioisotope tracer I^{131} to measure thyroid activity. The medical subjects were Alaska Native men and women in a number of villages of northern and central Alaska, and Air Force and Army servicemen.

In the research, capsules of radioisotope I^{131} were administered to Inupiat Eskimos, Athabascan Indians, and U.S. Air Force and Army personnel. The levels of radioiodide uptake in subjects' thyroid, blood, urine, and saliva was measured by scintillation (radioactivity) counters in the field and laboratory. Additional information was obtained through clinical examinations and by measuring blood cholesterol, dietary iodine uptake, and basal metabolism. To evaluate the effects of different levels of cold exposure, the study was conducted in different seasons of the year for Alaska Native subjects, and before and after exposure to cold stress with U.S. Air Force and Army subjects. The goal was to provide a picture of thyroid activity and its relationship to metabolism in the different racial groups exposed to cold. This study included six different tests and lasted from August 1955 to February 1957.

Dr. Rodahl received standard training on how to handle and administer I^{131}, and how to measure its uptake, retention, and excretion in humans. After obtaining the appropriate laboratory equipment for the project and approval from the Atomic Energy Commission for the use of the radioisotope, he instructed Dr. Gisle Bäng and other AAL physicians in the experimental field and laboratory procedures. A total of 200 doses were distributed in preformulated pharmaceutical capsules. The stated maximum doses of I^{131} for individuals given was 65 microcuries.[4] However, the usual dose was less, ranging from 18 to 50 microcuries, with most being 50 microcuries as was standard for radioiodine tracer studies of the time and as approved by the Atomic Energy Commission. Doses below 50 microcuries occurred, however, because of the natural reduction of radioactivity in the prepared capsules that occurred during the long transport time involved in shipping them to Alaska. (The researchers attempted to compensate for the lower does by using longer scanning times in the field, but those results were judged to be unreliable.) Although relatively good records of the study, including names of most subjects, were kept, some data were incomplete. Records of actual dosage for each subject during each trial were provided inconsistently in the physician's report.

CONDUCT OF THE STUDY: SAMPLE SIZE AND DISTRIBUTION

Alaska Natives

The first I^{131} tests took place in the coastal Inupiat villages of Wainwright and Point Lay during August 1955. The inland Inupiat village of Anaktuvuk Pass was visited in September, and the Athabascan Indian villages of Fort Yukon and Arctic Village were visited in October 1955 (see Box 1.2). Later in the study, AAL physicians made return visits to most villages and tested 46 subjects again to compare seasonal differences in I^{131} uptake. AAL doctors revisited Wainwright in February 1956 and the other villages except for Point Lay, which was not

---

[4]The microcurie (μCi) is a unit reflecting the level of radioactivity. One curie is equal to 37 billion disintegrations per second; 1 μCi is one millionth of a curie and is equal to 37,000 disintegrations per second. Higher μCi values indicate greater radioactivity.

Box 1.2
ALASKA NATIVES

The term "Alaska Natives" is used in this report to refer to the aboriginal people of Alaska as a whole, including Yup'ik, Inuit Eskimos, Indians, and Aleuts. The specific groups of Inuits included in the AAL studies are known as Inupiats, and the specific Indians are known as Gwich'in.

When the AAL physicians arrived at the villages selected for the study, they found mostly fixed sod or wood frame construction houses (Rodahl and Bang, 1957), though many people in the study from Anaktuvuk Pass and Arctic Village also lived in skin tents and were nomadic. The villages of Wainwright, Point Lay, Point Hope, Fort Yukon, and Anaktuvuk Pass, although made up of kinship groups (Spencer, 1984) rather than "tribes," had been reorganized under the Indian Reorganization Act of 1934 and were governed by village councils elected by the people. In every village there were also respected individuals, usually village elders, who had no elected position, but whose opinions were important to a majority of the people (VanStone, 1956). At least a few of the people in those villages spoke English at the time of the AAL study, and VanStone reported that at Point Hope the tendency was to elect some members of the council who could speak English and act as intermediaries in contact situations with whites. Arctic Village at the time of the study was a very small settlement and among the last to make contact with white civilization. An elder in the group of families that traveled together served as a leader, although he spoke no English (P. Stern and F. Newman, oral communication, 1994).

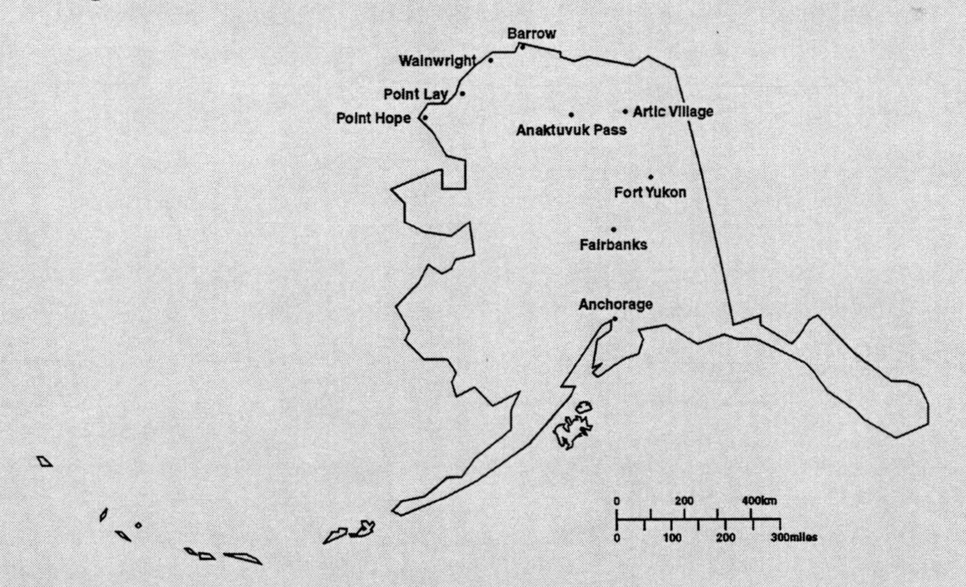

FIGURE 1.2. Locations of Alaska Native villages in which AAL experiments took place and other Alaska cities for geographic reference.

revisited, in March 1956. A third trip was made in July 1956 to Anaktuvuk Pass and Arctic Village, which were involved in a control experiment on potassium iodide uptake. Point Hope, an Inupiat coastal village, was visited only once late in the study, in February 1957.

A total of 76 men and 26 women, including women of childbearing age and some women who were lactating and one who was possibly pregnant, were subjects of the study. Ages of Alaska Natives in the study ranged from 16 years to 90. Table 1.1 provides a breakdown on the numbers of subjects, their gender and age distributions, and maximum (cumulative) radioactivity administered.[5]

Army and Air Force Servicemen

A total of 19 white servicemen were subjects in the study. Nine airmen and ten infantrymen were recruited from the military units in Fairbanks (K. Rodahl, personal communication, 1994). The first doses of radioisotope were given in September 1955 to four infantrymen and in October 1955 to four airmen. Measurement of thyroid uptake of radioiodide took place over a one-week period. Another group of five airmen was tested in the fall of 1955, and then again in February of 1956 following one month of field experience in the cold in unheated shelters. The second test of the first eight subjects took place in February and March of 1956. A third group of six infantrymen were given similar tests in October 1956 and February 1957 as a comparison sample to the Point Hope subjects.

The U.S. Air Force subjects were all males and ranged in age from 19 to 37. The maximum activity of I^{131} they received was 36 microcuries. For the Army infantrymen, the age range was 19 to 26, and the maximum activity of I^{131} given was 125 microcuries—the total dosages received by the six unidentified soldiers from the last test of 1956-1957. Table 1.2 provides basic information about the servicemen, their ages, and maximum activity of I^{131} received.

CONDUCT OF THE STUDY: SUBJECT SELECTION

The U.S. Air Force did not have guidelines in place requiring written consent[6] in human experimentation, so AAL physicians did not obtain signed consent forms. According to Dr. Rodahl (written communication, 1994):

---

[5]AAL Technical Report 57-36 (Rodahl and Bang, 1957, p.2) states that there were 120 participants—"19 whites, 84 Eskimos, and 17 Indians." This does not match the counts in Tables I, II, III, IV, V, and XXIII of that report showing test results from 121 participants.

[6]Information included in this section is based on testimonial information, correspondence, scientific reports of the time, and personal interviews. A more detailed evaluation of consent issues appears in Chapter 3.

TABLE 1.1. Alaska Native subjects of the AAL thyroid function study by village

VILLAGE	Number of Subjects	Number of Men	Number of Women	Age Range	Number of AAL Trips to Village	Highest I^{131} Activity in μCi
Wainwright	47	39	8	20-70	2	71
Point Lay	12	7	5	16-90	1	16
Anaktuvuk Pass	20	13	7	21-76	3	129
Fort Yukon	6	6	0	21-40	2	58
Arctic Village	11	5	6	17-67	3	
						130
Point Hope	6	6	0	19-28	1	54
TOTAL	102	76	26	16-90		

SOURCE: Rodahl and Bäng (1957).

TABLE 1.2. Military subjects of the AAL thyroid function study

Service	Number of Subjects	Age Range	Number of Tests per Person	Highest I^{131} Activity in μCi
U.S. Air Force	9	19-37	2	36
U.S. Army	10	19-26	2	125
TOTAL	19	19-37		

SOURCE: Rodahl and Bäng (1957).

> At the time of these studies [1955-1957] there was, as far as I know, no question of written consent. As a matter of fact, a number of our [N]ative subjects could neither read nor write. In the case of the [N]atives, we went to the leader of the group in question. . . and explained our proposed study and its purpose to him. He then talked to his group and came back to us with the subjects who had volunteered to take part in the study. Our studies were known to our medical colleagues, both military and civilian, and members of the Alaska Native Service . . . .In several cases they helped us with the initial contact with some of the [N]ative groups. Furthermore, our studies and the reports of our results were approved by the Head of the Arctic Aeromedical Laboratory . . .

The one military participant in the AAL study who spoke to the Committee did not recall signing a written consent form, although he did recall being briefed on the project and being asked if he wished to participate.

Generally, in the Alaska Native villages studied researchers approached village elders, explained the proposed research, and asked for help securing volunteers. Medical screening, basic care, and supplies were sometimes provided. Physicians gave the elders information about what they wanted to do and explained their need for volunteers. In turn, the elders recruited potential subjects. The explanations given by elders to potential volunteers in this recruitment process apparently were unwitnessed by the researchers. The elders' capacity to understand English and scientific language varied considerably from village to village. Thus, the information conveyed to subjects could have varied widely. Because there is no term for "radiation" in Alaska Native languages, it is not clear that full and accurate information about the study could have been conveyed to non-English-speaking subjects in any event. It does appear that Dr. Rodahl and colleagues communicated directly with subjects during the testing, at times through interpreters, to explain such things as what the subjects should do and when they should come back for more tests. He stated that the subjects who were brought to him were given the right to refuse to participate and that no one refused.

This approach to working with Alaska Natives is consistent with other AAL practices of the 1950s (Rodahl, 1962, p. 30; Hopkins et al., 1958, p. 122), in which a local elder was used to brief villagers or the village council on the research and give instructions. Local messengers would bring villagers to the makeshift clinic. In return the AAL physician would also run a clinic to treat illnesses in the local population. Results of the experiments generally were not conveyed to the subjects, only to the scientific press and colleagues, which presumably also included the Alaska Native Service doctors.

Wainwright

In Wainwright, a villager who is now an elder reported that he acted as the intermediary and interpreter for the AAL doctor in the 1950s medical study. During that period, Wainwright was a village undergoing a process of acculturation and had about 225 people who lived in 35 sod and wood frame houses (Milan, 1962). The elder (who was under 40 years of age at the time) was the chair of the village council and spoke English as a second language. The elder had obtained a vocational school education in Oregon in the 1940s and told the Committee that the physicians did not speak Inupiat. When the doctor said he wished to conduct a study and needed individuals from the village to participate, the elder discussed this request with the village council, which agreed to allow the AAL physician to conduct his study. It is unclear whether the elder received a complete explanation of the nature of the study, its potential risks, or the use of radiation in the study. He stated that the word "radiation" was never mentioned in connection with the use of the I^{131} capsules, and that he thought the capsules were to improve the villagers' health.

Individuals[7] chosen by the village council in Wainwright were told by messengers in the village to come to a schoolhouse where they were to see physicians. None of the eight participants[8] to whom the Committee spoke could remember receiving an explanation of the experiment, or being told they could refuse to participate. The village participants were asked about their ability to speak English at the time; while some said they had a rudimentary knowledge of the language from the local school, others said they did not.

Anaktuvuk Pass

The inland village of Anaktuvuk Pass was a community of 78 nomadic Inupiat Eskimos living in a collection of tents and nine houses who followed game, principally caribou, across the Brooks Range mountains and elsewhere in the interior of north central Alaska (Rodahl and Bäng, 1957; Laughlin, 1957). In Anaktuvuk Pass (and later Arctic Village), the AAL doctors encountered a health problem that had an impact on their research: endemic goiter. Goiter is indicated by swelling from an enlarged thyroid at the base of the neck. It is caused by insufficient intake of iodine and it is readily solved in our society by the use of iodized salt (see Box 1.3). The AAL doctor who visited the village reported that "16 of 27 persons above the age of 17 had definite thyroid enlargement." In mid-1956, the AAL doctor reported that three residents of the village were taken to Anchorage for thyroidectomy operations for goiter "since 1955" (Rodahl, 1957) or "by 1955" (Rodahl and Bäng, 1956b).

According to the five study participants from the village with whom the Committee spoke, the occurrence of goiter in the community was used in the description of the procedure by the AAL physician when the medical study was explained to the participants in the 1950s; however, the term "radiation" was not explained in connection with the experiment.

Some people from the village did not participate in the initial tests of September 1955 (it was said that they, or their families, were out hunting). Yet when the physician came back in March 1956, these people were included as subjects in the subsequent test, despite the fact that they would not have provided a before and after winter exposure for the purpose of the thyroid study.

In contrast to Wainwright, one participant from Anaktuvuk Pass recalled being offered the opportunity to refuse to participate, but agreed to the study because he believed he would benefit from doing so. Dr. Bäng recalled that he had informed an elder at Anaktuvuk Pass and

---

[7]One woman became part of the study despite the fact that she had an enlarged thyroid at the time. Although not given a subject number for further trials, she was given I^{131} apparently to study its uptake in a diseased thyroid. She was unable to recall the study, but told us she had a thyroidectomy in possibly 1959, four years after the physicians came to Wainwright. The Committee did not attempt to verify her medical history.

[8]The North Slope Borough's Department of Health and Social Services provided information that 20 of the 48 participants from Wainwright were still alive. For the other villages they said that 4 of 12 subjects from Point Lay were still alive, and 8 of 20 subjects from Anaktuvuk Pass were still alive.

Box 1.3
GOITER AND RADIOIODINE

When nutritional supplies of iodine are inadequate, or when there are defects in glandular processes, the thyroid sometimes swells to many times its normal size. When this happens it can produce choking sensations and difficulty in swallowing and breathing. Imbalances associated with inadequate production of thyroid hormones can have severe effects on human metabolism and cause severe mental retardation in the newborn (cretinism). While in its early stages, endemic goiter (goiter caused by inadequate supplies of iodine, often found in inland or mountainous regions of a country) can be treated by supplementing the diet with salt containing potassium iodide. More severe enlargement may require surgery for partial or complete removal of the gland (thyroidectomy).

Based on the knowledge that iodine is concentrated in the human body's thyroid gland, researchers in 1937 began experiments on use of radioactive isotopes of iodine to trace the activity of the thyroid gland. Using I^{130} initially with radioactivity measured by a geiger counter, physicians were able to trace the accumulation of the radioactively tagged iodine in the thyroid and bloodstream and rate of removal from the body in urine (Chapman, 1983).

Experiments in 1943 determined that large dosages of radioiodine in millicurie amounts could destroy enlarged thyroid tissue, eliminating or reducing the need for surgery. In 1946, the U.S. Atomic Energy Commission was allowed to produce I^{131} (Chapman, 1983). From that time on, radioiodine became a widely used diagnostic and therapeutic agent. Development of the scintillometer allowed a reduction in the amount of I^{131} needed for diagnostic testing. While concerns over potential cancer causation (Rall, 1956) resulted in the use of less potent isotopes of radioiodine, I^{131} is still used now as a therapeutic agent for thyroid disease.

the director of the private hospital in Fort Yukon about the study and explained the purpose to the participants, who were selected "on a voluntary basis" (written communication, 1994).

Fort Yukon

Fort Yukon, an interior Alaskan village of Athabaskan Indians with a population at the time of between 400 and 500, was the location of a private hospital (Rodahl and Bäng, 1957, pp. 20-21). At that location in October 1955, six men were selected for the study. Some participants were employed at the Fort Yukon Hospital at the time; some, in fact, participated in previous and subsequent experiments by the AAL doctors (R. Balaam, oral communication, 1994). As in visits to the other villages, the word "radiation" was not used by the AAL doctor in connection with the experiment. The participant with whom the Committee spoke said he spoke English at the time and understood the term, but it was not mentioned; he said he took part because he was curious about what they might do.

Arctic Village

During October 1955 and March and July 1956, AAL doctors visited Arctic Village. At the time, the village was a part-time settlement of families of Athabaskan Indians in the interior of Alaska who mostly were subsistence hunters living off caribou and other game (Rodahl and Bäng, 1957; Rodahl, 1956). The village was one of the very last to be visited by white civilization, and at the time no one spoke English. The families traveling together were approached by the AAL physician, who landed by small airplane, but it is not known how communication was established or information provided on the thyroid test. Eleven villagers were tested, although not all were present when the doctor arrived each time. The daughter and son of the village chief told the Committee that they spoke no English at the time and all they could do was nod at the doctor. Rodahl and Bäng (1956a) reported that in April 1955, four cases of goiter had been found in the population, and by mid-1956, 9 of the 50 Indians had enlarged thyroids. Two Indians had been transported to Anchorage for thyroidectomies since 1955 (Rodahl and Bäng, 1957).

Arctic Village and Anaktuvuk Pass were chosen by the AAL doctors for a control test using a potassium iodide supplement to determine whether increased I^{131} uptake was related to endemic goiter or might be related to thyroid stimulation from exposure to cold. In the test, four of the subjects from Anaktuvuk Pass were given 0.6 milliliters of potassium iodide daily for three months, after which their thyroid uptake of I^{131} was tested; five other village members were not given the supplement before their retest. In Arctic Village, two subjects were given the supplement after their first thyroid test and four others were not. The subjects do not recall being told why they were given the supplement, but the supplement ended after the doctors completed their control studies in 1956. The subjects receiving the dietary supplement showed a significant drop in iodine uptake in the follow-up I^{131} testing. The doctors also performed a nutritional analysis for Anaktuvuk Pass and Arctic Village and found an inadequate intake of iodide in the salt, food, and water supply. On the basis of this work, they concluded "that the observed deviation from the normal in the iodine metabolism in the inland Eskimos and Arctic Village Indians is a manifestation of endemic goiter and cannot be taken as an indication of the effect of cold exposure on thyroid function" (Rodahl and Bäng, 1957), the chief reasons being lack of access to marine food and the use of uniodized salt (Rodahl and Bäng, 1956a).

Point Hope

In February 1957, six men from Point Hope were tested for their thyroid iodine uptake to see why after 24 hours after the tracer dose, blood content of I^{131} was lower in Natives than in whites and lower in the winter than in the summer. The subjects were tested once in comparison with a group of six infantrymen. Unfortunately, the names of the 12 subjects were not recorded and the Committee was unable to locate any AAL archive records that might have included such data. A semiannual report of the AAL (U.S. Air Force, 1956b) stated that Dr. Bäng was the only AAL physician who visited Point Hope in 1956; the same report also mentioned the selection of test subjects for the 1957 thyroid study. The Committee could not tell who tested the infantrymen, except the Committee does know that Dr. Rodahl had left Alaska by that time.

Military Testing

In September to October 1955, volunteers were sought by Dr. Rodahl from the Air Force and Army enlisted men stationed in Fairbanks. The method used, according to Dr. Rodahl, was a request to the commanding officers, who then made an announcement requesting volunteers. Volunteers were told the nature of the experiment. According to the report, four Army infantrymen and nine Airmen, including four volunteers from the AAL, were tested initially for I^{131} uptake including saliva, urine, and blood samples over four days after basic weight, height, age, pulse, and blood temperature readings were made. In a follow-up, the volunteers were tested again during February and March 1956. In addition, volunteers who were to spend one month in January 1956 on field maneuvers in the cold without benefit of warm shelter were tested for thyroid activity before and after the field maneuvers. The one participant with whom the Committee spoke did not remember specific details about how he was informed of the experiment, nor did he remember taking the iodine tracer capsules. He did remember the cold weather maneuvers, in which he stayed outside for a winter month near Fairbanks with only unheated shelter; during that period he also participated in other experiments related to survival in cold weather. A second experiment using six (not named) infantrymen took place in the months of October 1956 and February 1957.

STUDY RESULTS

The AAL thyroid function study concluded there was no significant difference in thyroid uptake or urinary elimination of radioactive tracers in the Natives and whites tested. Nor was there any indication of increased thyroid stimulation as a result of exposure of the subjects to cold.

The AAL study (Rodahl and Bäng, 1957) concluded there was no racial difference in the uptake or excretion of I^{131} and protein-bound I^{131}, and no consistent seasonal difference either. The increased uptake in I^{131} in inland Inupiat Eskimos and mountain Athabascan Indians was attributed to iodide deficiency in the diet in these populations. As a result of the winter field exercises for military subjects, no evidence was found of increased thyroid stimulation. The thyroid, the doctors concluded, did not play a significant role in human acclimatization to the Arctic environment.

The military continued to investigate the role of the thyroid in acclimatization to cold. In Alaska in the 1960s, at least one more AAL study using military volunteers, and possibly another by the Naval Research Laboratory,[9] was conducted. Over time, some relation was discovered, but not until techniques were improved. As described in Appendix A "Thyroid Function in Health and Disease," subsequent research identified what was named the "polar T_3 syndrome" (Reed et al., 1990). In white men who lived in Antarctica for more than five continuous months, Reed, Brice, and colleagues (1990a) described increased production of thyroid hormone T_3, as well as increased metabolic clearance and increased serum binding indicating a more efficient use of the T_3 hormone in the body in response to cold. Another discovery of thyroid-related response to lengthening daylight correlated production by the pituitary of thyroid-stimulating hormone (TSH) and cholesterol in the blood (Tkachev et al., 1991). (Levels of TSH and the thyroid hormone T_3 in the blood increased while levels of hormone T_4 and cholesterol decreased with the lengthening of daylight duration.)

A review of data from tables in the Rodahl and Bäng study (1957, pp. 49, 73) demonstrated the T_3 effect, in that retention of the I^{131} was greater in human subjects after exposure to winter climate. The experiments conducted from 1955 to 1957 concentrated on examining racial differences of iodine uptake in the thyroid. By not noting the decreased excretion of iodine in their winter experiments, the physicians failed to understand that the human body was more efficiently using thyroid hormone. Blood analysis techniques were not sophisticated enough at that time to measure the endocrine secretion levels.

THE COMMITTEE'S METHODS

The experiments at issue here happened 40 years ago, so reconstructing the events of the research and locating the researchers and subjects was a formidable task. The effort consisted of a detailed historical records search, requests for assistance from local, tribal, state, and federal agencies and governments, telephone and personal interviews, written interviews utilizing a questionnaire, and a public hearing. The Committee focused on gathering information on the conduct of the medical study and did not conduct physical examinations of the study participants nor review their medical histories, although that information was volunteered in the testimony of some participants. The Committee did attempt to contact all subjects of the study, but address changes, the death of some participants, and lack of pertinent information either in the original AAL report or in Air Force records prevented the Committee from locating all participants. At times, the Committee had to rely on anecdotal evidence because this was all that was available.

---

[8]A report of a meeting found in the archives of the University of Alaska, Fairbanks, provided details of an experimental protocol to be carried out by the Naval Arctic Research Laboratory in Barrow from November 1967 to February 1968. The study would have examined uptake, retention, and excretion of I^{125} after exposure to cold weather by 12 Wainwright Natives. The Committee could not determine, however, whether the study was ever carried out.

Written Information

To conduct this evaluation, the Committee needed as complete a written record as possible of the experimental procedures. Documents used included copies of technical reports of the AAL, semiannual histories of AAL activities, and approval documents by the Atomic Energy Commission for the conduct of the AAL research. Using those materials as a start, the Committee requested a search by the Air Force and the National Archives for AAL internal approvals for experimental design and conduct of the experiment, memorandums, reports, or orders for the physicians and support personnel to fan out across Alaska to conduct the study, orders or voluntary assignments and personal waivers for enlisted personnel to participate in the study, and interim and final reports to Air Force personnel from the period of the study.

The records search was conducted in the archives and libraries at Brooks Air Force Base, San Antonio, Texas; Bolling Air Force Base, Washington, D.C.; Elmendorf Air Force Base, Anchorage, Alaska; Maxwell Air Force Base, Montgomery, Alabama; Wright-Patterson Air Force Base, Columbus, Ohio; and the Department of Defense records center in Alexandria, Virginia. The documents search also was conducted by archivists at the National Archives regional centers in Anchorage, Alaska; Seattle, Washington; St. Louis, Missouri; Washington, D.C.; and Suitland, Maryland. No records of the AAL prior to 1958 were located.

Interview and Public Hearing Information

The written documentation available was not adequate to provide a complete picture of the AAL thyroid function study, particularly in regard to the consent process, the day-to-day functioning of the AAL at the time, and the participants' understanding of the nature of the study (both researchers and subjects). To obtain further elaboration on the methodology and conduct of the experiments, the Committee sought personal accounts from participants in the research. The Committee had help from various sources in locating participants. Using information from AAL records, which listed all but 12 of the experimental subjects, the North Slope Borough's Mayor's Office identified subjects in villages under its jurisdiction and shared that information with the Committee. The Tanana Chiefs' tribal organization did the same for Athabascan Indian villages in central Alaska. Some information was also provided by the U.S. Indian Health Service.

On July 5, 1994, a member of the Committee (Dr. Baines) and the NRC study director met and interviewed eight Alaska Native participants of the study in Wainwright, Alaska, and a village elder who had acted as the intermediary for the AAL physicians; three of the eight did not remember the research. On July 6, they interviewed six Alaska Native participants of the study in Anaktuvuk Pass (one did not remember the research). In the trips to the two villages, the North Slope Borough's Mayor's Office provided an Inupiat translator with a medical background to assist in translating questions and responses from subjects who spoke no or limited English, and a special assistant who knew the house locations of the experimental subjects. Two subjects from Fort Yukon were reached by telephone, but could not remember the details of their participation.

The Committee then hosted a two-day public information-gathering session, July 7 and 8, 1994, to gather first-hand information from as many participants as possible. Several witnesses were identified in advance and asked to speak during the session. Because of time and distance constraints, some witnesses unable to travel to Fairbanks provided their testimony by telephone conference call. Interested members of the public were also invited to provide information. Although not intended as an epidemiological follow-up, the Committee used a mailed questionnaire to seek responses about the conduct of the experiment from the known subjects who were not interviewed in their villages, over the phone, or at the public session by the Committee.

Speakers knowledgeable on the subject of medical practice and health concerns in Alaska in the 1950s and the 1990s also were identified and asked to make presentations on State of Alaska health concerns and programs (of the 1950s or 1990s or both, depending on expertise), scientific and ethical concerns regarding human medical experimentation of Alaskans and military men during the period of the Cold War, and the I^{131} studies in particular.

Some former employees of the AAL during the period 1955-1957 were asked to speak on the working conditions and organization of lab and about the state of medical practice in local communities, the relationship between the Alaska Native Service and the staff of the AAL, methods used to get military and Native volunteers for medical studies, what guidance on informed consent may have existed in the AAL at that time, and what they could remember about the AAL studies. This background information helped the Committee understand the day-to-day function of the AAL in the 1950s time period.

Subjects of the experiments were asked a standard list of questions about how the study was done; how they were picked to be in the study; what they were told by the Air Force doctors before, during, and after the study; if the doctors spoke with village elders or the witnesses (or their parents if they were very young) to get consent to do the study and what they said to them; if they were told there were any possible ill effects that might be suffered from participating in the study; and if the researchers or other doctors from the Air Force or the Alaska Native Service saw the subjects after the study was completed or gave any medical care as a result of the subjects' taking part in their study. These questions were included in a letter sent to each potential witness and an announcement sent to the Inupiat villages. In addition, a public announcement of the public session was provided to press sources in Alaska to seek additional witnesses who might come forward to offer relevant testimony.

The Office of the Air Force Surgeon General was able to provide addresses for a very few of the military experimental subjects; NRC staff sent these individuals a letter and attempted to reach them by follow-up phone call. Unfortunately, only one of the 13 named in the AAL study report could be reached. This person participated by phone and was asked basically the same questions as the Native participants, except he was also asked about the process of military volunteering and consent. Information was gathered from Dr. Rodahl and Dr. Bäng via telephone calls, correspondence, and a telephone conference call with the Committee during the public session. They answered questions on all phases of the 1950s experiments. Details from the public session appear in Appendix B.

COMMON THEMES FROM THE PUBLIC SESSION

The public hearing and interviews were free of obvious rancor and reflected a real willingness to cooperate and share information with the Committee. All the participants, but especially the Alaska Natives, were exceedingly generous in their attitude toward the Committee's mission and their willingness to be of help, even at great personal expense (financial and emotional). The advantage of such interchange was that it permitted the Committee to gain insight into what really occurred and what the participants thought and felt. The Committee heard the following common themes during the public session:

Participants in the AAL thyroid function study and the public in Alaska share a general confusion about the I^{131} *experiment and other medical experiments conducted in the region through time, as well as confusion about a host of Arctic contamination issues.*

During the public session, it became clear that many of the speakers were concerned about experiments beyond the AAL thyroid function study—other medical experiments conducted at other times and places. Some speakers raised concerns about unrelated, but obviously important, issues such as unresolved Arctic contamination from the military presence. Some spoke with alarm about reports of radioactivity in the Arctic Ocean from dumping waste materials such as sunken nuclear reactors from the former Soviet Union. Questions were posed about other radioactive elements besides I^{131}, including strontium and cesium.[10] For some, the public session was seen as an opportunity to air grievances that had no forum before.

Participants felt fear, anger, and uncertainty about the role of I^{131} *and other contaminants in producing various chronic ailments and even death.*

Some of the Alaska Natives who testified believed that Native people seemed healthier long ago and that health difficulties seemed linked with the rapid and sometimes dramatic changes caused by increasing domination by Western culture that began in the 1940s. Again, this concern is not directly related to the AAL thyroid function study, but is important to our understanding of the people's feelings about the study. It is impossible to disentangle completely the people's reactions to other research that was conducted and other aspects of their contact with the incoming dominant culture and the changes caused in their traditional lifestyle.

Participants in the public session expressed great frustration and increasing militancy about the lack of respect given Native people by the Western majority. Throughout the public session, speakers voiced frustration that Native people repeatedly had been betrayed by white authority figures and that these betrayals frequently followed extensions of hospitality, respect, and trust by Native peoples. As evidence of disrespect, speakers cited the obvious indifference and ignorance of whites about the Native government and social structures, and about Native philosophy of life. White arrogance was characterized by a persistent presumption that whites believed that Natives "wanted to be like us." The tension between the cultures was clear when one participant asked the AAL principal investigator if he thought Natives were human. Some

---

[10]Atmospheric testing of atomic weapons by the U.S. and the former Soviet Union in the 1950s and 1960s resulted in some accumulation of nuclear fallout in Alaska. This matter was still brought up as a problem by testifiers at the Committee's hearing. Although this was not the focus of the Committee's efforts, several studies such as those of Hanson (1967, 1971, 1982, 1994), and Stutzman et al. (1986) have addressed this matter and found that the fallout did not result in cancers in local Alaska populations.

speakers, and even the Committee, were chastised for being insufficiently attuned to, or carelessly informed about, how different were the needs and concerns that distinguished the Inupiat Eskimos from the Indians. Participants pointed to who benefitted from the research as another example of white arrogance—the experiments on Natives were performed not to help the subjects or to help their general society, but to help the U.S. military and the general welfare of whites. When given this opportunity to voice long-held frustrations, a theme was that the research epitomized white entitlement at the expense of Native respect and trust.

Participants doubted the willingness of the U.S. government to share information and wondered whether even this attempt at openness was half-hearted.

With hindsight, the secrecy ethos of the Cold War era did much to breed mistrust of government, not just in Alaska but throughout the citizens of the nation. The Alaska Natives' strong response to learning of the AAL thyroid function study was born out of years of mistrust, a mistrust that will not be simple to rectify. To the interested parties it may have seemed peculiar that there is not a larger and clearer paper trail about the research subjects. Participants in the public session expressed frustration at how difficult it has been for them to obtain archived information. Even this Committee was suspect—was it part of an overall "stonewalling" effort because it was unable to supply names of some individuals who participated in the I^{131} experiments? Further, some were dissatisfied because of the short notice given about the Committee's public meeting; this problem was due to time constraints but could be interpreted as an indication of insincerity. Participants also were distressed that the Committee did not visit every concerned village and that not all members had visited villages; this constraint was also due to time and financial limitations.

Over and over again, speakers voiced concern about the history of abuse the indigenous people have suffered from the dominating mainland culture. From the Natives' viewpoint, their grievances have been met with at best patronizing indifference and at worst lethal disregard. Although not specific to the AAL thyroid function study under investigation by this Committee, these concerns are reactions to the larger context within which the study took place, a context which has implications for the nature of the conduct of the study.

2

Health Effects of I^{131} Administration in Humans

Small amounts of radioactive materials are often used for both observation and diagnosis of body systems and treatment of disease (see Box 2.1). The radioactive tracer iodine 131 (I^{131}) has been used to evaluate thyroid function in humans since the 1940s. In fact, the field of nuclear medicine developed primarily from the successful use of I^{131} as a probe to evaluate thyroid function. In the five decades since the introduction of I^{131}, thousands of radiopharmaceuticals have been developed to evaluate function and disease in almost every tissue of the body, and millions of radioiodide tests have been performed on Americans. Today, over 10 million nuclear medicine examinations are performed annually in the United States (NCRP, 1989).

I^{131} was the only radioactive tracer readily available for use in the 1950s, when the AAL thyroid study was conducted. However, serious shortcomings limit its use in modern nuclear medicine. The major drawback of I^{131} is the relatively high radiation dose received by the thyroid. Assuming an uptake of 25 percent of the administered radiation activity, the thyroid receives a dose of approximately 1.3 rad per microcurie[1] of administered radioactivity. Another disadvantage is that principal gamma ray emissions from the radioisotope are not efficient for imaging purposes; the 364-keV photon emitted is higher than optimal for gamma scintillation cameras currently used in nuclear medicine. Today, the radioisotopes technicium 99m and I^{123} are preferred in thyroid imaging procedures except when diagnosing thyroid cancer, in which case other nonradiogenic (ultrasound, palpation, or thin-needle aspiration) means may yield useful information. These radionuclides have short half-lives and decay emissions, which reduce the radiation dose to the thyroid. Their gamma energies (in the range of 150 keV) are also optimum for imaging studies.

---

[1]See Boxes 2.1 and 2.2 for background on radioactivity and how radiation affects the human body.

RADIOLOGICAL BACKGROUND OF THE AAL STUDY

The use of radioactive materials at the AAL for animal and human research studies was granted by license from the U.S. Atomic Energy Commission (AEC).[2] The license allowed the use of I^{131} in humans to determine uptake of iodide in human subjects before and after cold exposure. The standard procedure involved administration of I^{131} in approximately 50 microcurie doses. Some subjects received multiple doses at one- to six-month intervals. The AEC license required that the investigator be properly trained in radioisotope usage and methods in order to procure, use, and dispose of radioactive materials in research studies. Dr. Rodahl, the principal physician and investigator on the AAL project, had at the time of the experiments completed the required training, which involved 30 hours of participation in an instruction program and experience diagnosing and treating 49 cases using radioactive materials. Dr. Rodahl's training in the principles of radioactivity, radiological health safety, radiation measurement techniques, and biological effects was completed at the Cook County Graduate School of Medicine in Chicago.

The AEC license provided for the human administration of I^{131} in the form of capsules, which were to be preassayed by the supplier. The radioactive material for the AAL thyroid study was obtained from Oak Ridge National Laboratory in Tennessee, considered the most reliable supplier available. Each capsule contained approximately 50 microcuries of I^{131}. Because of the relatively short half-life of I^{131} (eight days) and long travel time involved, the supplier had to provide capsules with higher activity than this to account for radioactive decay during transportation of the material from Oak Ridge to the AAL in Fairbanks. For instance, if the transit time was one week from the time of manufacture at Oak Ridge to use in human subjects, the Oak Ridge facility would have had to formulate I^{131} capsules with an activity of 92 microcuries for the capsules to have 50 microcuries at the time of administration. In some cases, the dose administered was recorded as 65 microcuries, which may have been a result of early delivery of the capsules.

OVERVIEW OF EPIDEMIOLOGICAL EVIDENCE REGARDING RADIATION-INDUCED THYROID CANCER

Since 1990, there have been several comprehensive reviews of the epidemiological literature concerning radiation induction of thyroid cancer in humans (NAS, 1990; Shore, 1992; Ron et al., 1995). Studies can be categorized into the following groups: (l) patients with various medical conditions (including benign disease and cancer) who received external beam therapy (x-rays) in which the thyroid was incidentally exposed, (2) patients given I^{131} for medical diagnostic or therapeutic reasons, and (3) individuals exposed to nuclear radiation and/or fallout

---

[2]AEC licenses 46-50-1 and 33252, dated May 20, 1955.

Box 2.1
HOW DIAGNOSTIC AND THERAPEUTIC RADIATION AFFECTS THE HUMAN BODY

Biologists and physicians use radioactive materials for both observation and diagnosis of body systems and treatment of disease. Small amounts of radioactive materials called tracers can be used to see how certain chemicals are transported or are stored in the bodies of plants, animals, and humans. Some of the same radioisotopes—in substantially larger dosage amounts—can be used to treat cancer or disease. Both uses rely on radiation of emitted particles and rays from radioisotopes. Specific radioisotopes may concentrate in a particular organ or gland, and emit radiation that is absorbed by that organ or gland. This absorption occurs by a process in which the irradiated particles pass through tissue, exciting and ionizing atoms and molecules in their path. Most of the absorption takes place in water, since cells are made up of more than 70 percent water.

Various isotopes of an element are chemically identical and follow the same path and uptake in living tissue; radioisotopes have the particular characteristic of being detectable as they travel. As a radioisotope breaks down, it emits particles and rays that can be measured with radiation detecting equipment or specialized photographic equipment.

When substantially higher doses of radioactive materials are used to treat disease, a radioisotope that concentrates in a particular organ or gland will give off radiation that kills the cancerous cells. In this case, radiation is used to destroy living tissues—cancer cells that are undergoing uncontrollable division, which makes them particularly vulnerable to the effects of radiation. Cell death or damage occurs in two basic processes. In the first, electrical changes that take place in the water of a cell may create free radical molecules that attack molecules in the cell, leading to biological damage. In the second, direct interactions of irradiated particles at sites of a cell's DNA (the basic molecule that carries the genetic information in the cell for replication and carrying out life functions of the cell) cause damage to the cell and its ability to repair the damage or replicate itself.

Many factors determine the effect of radiation at the cellular level. While impact on DNA is the most important, also significant are the quality of the radiation (affected by dose, dose rate, and the type of living tissue involved), cell sensitivity to the radiation (which can vary as a function of its state in the cell cycle, the dose rate, the size of the dosage) and the presence of other body chemicals which may reduce or increase the effect of the radiation. Over a period of many years, called a latency period, damage at a localized site may result in the formation of cancerous cells-but this is dependent on the many factors listed above.

The intake of radioisotopes into the human body can be by inhalation, ingestion, injection, absorption through the skin and mucous membranes, or through cuts and abrasions. Each of these different exposure routes has its own pattern of deposition on or in parts of the body such as the lungs, gastrointestinal tract, or skin. While the radioisotope is found at one of these sites, the surrounding tissue is irradiated, and the extent of this irradiation will be determined by factors related to the dosage. Radiation effects differ between that administered from outside the body (e.g., x-rays, or atomic fallout) or internally (e.g., 1^{131} capsules). In the latter case, the radionuclide enters the body, but the dose it delivers to various organs and tissues varies and the radiation is uneven across the body because of the affinity of certain radioisotopes for certain organs or glands. As a result of this concentrating effect, epidemiological studies of the long-term effects of radiation exposure to specific isotopes focus on cancers of the target organ or gland, the most likely site for cell damage. In the case of the AAL study, the most likely site is the thyroid gland.

SOURCE: National Research Council, 1990

radiation (including radioactive isotopes of iodine) from nuclear weapons.[3] A number of generalizations about radiation-induced thyroid cancer may be drawn from these studies, and these conclusions can help provide a reasonable assessment of the radiation risk for thyroid cancer in the AAL study.

1. There are major differences in background thyroid cancer rates among the various cohorts studied, perhaps related to differences in lifestyle. The development of thyroid cancer from initiated cells (cells which have received some type of carcinogenic stimulus) is greatly dependent upon hormone status.

2. Females are two to three times more susceptible to radiogenic and nonradiogenic (i.e., background) thyroid cancer than males. The relative risks do not differ between the sexes.

3. Thyroid cancer risk from external irradiation has a significant age dependency. Risk is particularly high among children exposed within the first five years of life. There is little risk in

---

[3]Exposure to nuclear radiation or fallout from nuclear weapons brings substantially greater risks than incidental, diagnostic, or therapeutic exposure but the former does offer lessons of relevance.

Box 2.2
FUNDAMENTALS OF RADIOACTIVITY

Atoms, the building blocks of all solids, liquids, and gases, are joined into more complex structures called molecules by bonds formed between subatomic particles called electrons. Atoms contain a core, or nucleus, with smaller particles such as protons and neutrons inside. Radioactivity is caused by atoms emitting subatomic particles and high energy rays. When this radiation bombards cells, if the energy of the radiation is high enough, it can displace atomic electrons and break the bonds that hold a molecule together, changing, the chemistry of living cells. Radioactive substances are both naturally occurring and man-made. There are three basic kinds of radiation: alpha particles, beta particles, and gamma rays. Gamma rays are the highest energy radiation.

Radiation is emitted from the nuclei of atoms. Alpha or beta particles originate in the nucleus and their emission causes the original atom to change into an atom of another element. Gamma rays release energy only and do not cause such a change. The number of radioactive particles emitted over a period of time by a radioactive substance is directly related to a percentage of the number of atoms in the substance present. The period of time it takes for one-half of a radioactive substance (radioisotope) to change into another is called its "half-life"; different radioisotopes have different half-life periods that can range from fractions of a second to billions of years.

Alpha particles have a positive electrical charge and consist of two protons and two neutrons bound together; their energy is dissipated by passage through small amounts of matter (an alpha particle cannot pass through more than one or two sheets of paper, for example). Beta particles are electrons that may be negatively or positively charged, but can penetrate much farther into dense matter (13 millimeters into a piece of wood, for example). Gamma rays are high energy electromagnetic radiations similar to x-rays which can penetrate great distances through matter. X-rays are produced outside of the nucleus whereas gamma rays originate inside the nucleus.

individuals exposed after ages 15-20 years. The risk in adults (greater than age 15 years at time of exposure) may be one-eighth to one-tenth the risk in children.

4. Current epidemiological evidence on thyroid cancer induced by external irradiation is consistent with a linear dose-response curve. In children, there is convincing evidence for risk at about

10 rad. (The epidemiological data can be fit by a relative risk model because of the strong dependence of risk on the natural incidence of thyroid cancer in the population under study. However, an excess risk model fits the data best and includes allowances for cohort effects, latency, age at exposure, and sex.)

5. I^{131} is estimated to be 20-25 percent as effective as externally administered x-rays in producing thyroid cancer. Although the reason for this low carcinogenic potential is not well understood, it may be related to the facts that I^{131} is concentrated in the colloid, reducing the dose to the follicular cells, and that the dose rate from I^{131} radiation is reduced compared with external radiation exposures. Spreading the dose over time may reduce the risk because of cellular repair of radiation injury.

6. There is a characteristic time lag (referred to as the "latent period") between external radiation exposure and clinical appearance of thyroid cancer. The minimum latent period appears to be about five years. Risk remains elevated for 30-40 years after exposure.

Table 2.1 identifies several major studies that provide risk estimates of radiation-induced thyroid cancer (Ron et al., 1995). All the studies cited involved external irradiation of subjects as a result of atomic bomb irradiation or radiation therapy for the treatment of various benign diseases or cancer (other than thyroid cancer), rather than diagnostic levels, but information from these studies can be helpful in understanding thyroid cancer risk generally. Brief descriptions of these studies follow; more detailed information is available from the comprehensive reviews by Shore (1992) and Ron et al. (1995).

The largest single study of thyroid cancer risk involved Japanese survivors of the atomic bombings in 1945 (Ron et al., 1995). In this study, thyroid cancer incidence diagnosed between 1958 and 1985 was determined among 79,972 atomic bomb survivors. The average thyroid dose was 27 rad (range: 1-399 rad). This is the only study that includes people of both sexes who were exposed at all ages.

Radiation-induced thyroid cancer has been documented in children irradiated for enlarged thymus in a Rochester, New York study begun in the early 1950s (Shore, 1992; Ron et al., 1995). The study included 2,856 subjects treated between 1926 and 1957 and all 5,053 nonexposed available siblings. All patients were exposed before the age of 1 year. Average dose to the thyroid was 136 rad (range: 3-1,100 rad).

Radiation therapy has been used to treat ringworm of the scalp (tinea capitis), leading to incidental irradiation of the thyroid gland. One relevant Israeli study included 10,834 persons treated in this manner between 1948 and 1960, plus 10,834 disease-free, nonirradiated matched comparison subjects and 5,392 disease-free, nonirradiated siblings. All patients were treated before the age of 16 years. The average thyroid dose was estimated to be 9 rad (range: 4-50 rad). The risk of thyroid cancer in the Israeli study was determined to be substantially higher than in the other studies listed in Table 2.1. Whether the higher coefficient is due to statistical fluctuations, to unusual susceptibility within this population, or to other factors is unclear (Shore, 1992).

Thyroid cancer has been observed in children irradiated for benign head and neck conditions (primarily enlarged tonsils and adenoids). In one study, more than 5,000 patients received head and neck radiotherapy in Chicago between 1939 and 1962. Adequate medical follow-up and dose information could be obtained for 2,634 subjects. Thyroid gland doses ranged from 1 rad to 580 rad, with a mean dose of 59 rad. In another study from Boston, thyroid cancer incidence was compared in 1,590 irradiated children treated between 1938 and 1969 and 1,499 children who were treated by surgery only for removal of enlarged tonsils or adenoids. The mean estimated thyroid dose was 24 rad (range: 3-55 rad).

Thyroid cancer has also been observed in individuals who have been given radiation treatments for other cancers. Two studies provide limited thyroid cancer risk information. One study included 9,170 childhood cancer patients who survived two or more years. The average thyroid dose was substantial—1,250 rad (range: 100-7,600 rad). In a large international cohort study involving 150,000 cervical cancer patients, 43 women were identified who developed thyroid cancer at least five years after their diagnosis for cervical cancer. Eighty-one controls were matched individually to these cases. The mean thyroid doses was 11 rad (range: 1-24 rad). This is one of only a few studies which documents radiation-induced thyroid cancer in adults.

Combined, the studies discussed above include about 120,000 people, nearly 700 thyroid cancers, and about 3 million person-years of follow-up (Ron et al., 1995). Except for the atomic bomb survivor and the cervical cancer patient studies, which include adult subjects, risk estimates provided in Table 2.1 are based primarily on observations made in children. For childhood exposures to external radiation, the pooled absolute risk is 4.4 per million person-years per rad (Ron et al., 1995).

Although our understanding of thyroid cancer risk is based primarily on studies of children exposed to external radiation, there are several epidemiological studies that have explored the effects of I^{131}. Table 2.2 lists several epidemiological studies of I^{131} exposures in children and adults and estimates for thyroid cancer risk. Shore (1992) provides detailed discussions of these studies in the context of an overall review and analysis of radiation-induced thyroid cancer epidemiology. Comparison of risk estimates in Tables 2.1 and 2.2 indicate that exposure to I^{131} is not as detrimental as external radiation exposure.

A few epidemiological studies of diagnostic I^{131} have been conducted (the Swedish diagnostic study, the Food and Drug Administration (FDA) study, and the German diagnostic study). These studies are particularly relevant to the AAL study because they involved diagnostic levels of I^{131} in which radiation doses to the thyroid are comparable (Table 2.2). None of these studies provide clear evidence of excess thyroid cancer. The largest of these investigations was conducted by Holm and his colleagues (1988), who studied retrospectively some 35,000 patients given diagnostic doses of I^{131} in Sweden between 1951 and 1969, and followed up for an average period of 20 years. The average activity administered was 52 microcuries of I^{131}, levels similar to those used in the AAL study. The mean age at the time of I^{131} use was 44 years; 5 percent of the subjects were under age 20. Increased risk of thyroid cancer was observed only in patients originally examined because a thyroid tumor was suspected. Patients given I^{131} for reasons other than a suspected tumor were not found to be at increased risk. There was no significant excess of thyroid cancer in this study, which had good statistical power to detect effects in adults (Shore, 1992). Shore (1992) noted that the carcinogenic potential of I^{131} beta particles may be as low as one-fifth that of external x-rays or gamma rays.

TABLE 2.1. Risk of thyroid cancer after external radiation exposure

Study Population	Excess Absolute Risk[a] (10^{-6} PY-rad)	Excess Relative Risk (per 100 rad)
Exposure at <15 years old		
Thymus	2.6 (1.7, 3.6)[1]	9.1 (3.6, 28.8)
A-bomb	2.7 (1.2, 4.6)	4.7 (1.7, 10.9)
Tinea capitis	7.6 (2.7, 13.0)	32.5 (14.0, 57.1)
Tonsil	3.0 (0.5, 17.1)	2.5 (0.6, 26.0)
Childhood cancer	ND	1.1 (0.4, 29.4)
Exposure at ≥15 years old		
Cervical cancer	ND	34.9 (-2.2, 00)
A-bomb	0.4 (-0.1, 1.4)	0.4 (-0.1, 1.2)

SOURCE: Ron et al. (1995)
[a]95 percent confidence limits in parentheses.
ND = not determined.

CALCULATIONS OF RADIATION RISK

Risks of radiation-induced thyroid cancer have been estimated in Table 2.3. The risk (probability of radiation-induced thyroid cancer) is the product of the radiation dose to the thyroid and the absolute risk coefficient (excess number of cancers per million persons per rad). For each population group, the average thyroid dose was determined by multiplying the average activity administered by a dose conversion factor. The average activity noted for each population group was derived from the AAL reports; the dose conversion factors (which convert the amount of radioactivity administered into a thyroid radiation dose) were based on calculations provided by Oak Ridge Institute for Science and Education based on current computer models of internal radiation dosimetry of the thyroid gland for I^{131}.

An absolute risk coefficient of 4.4 x 10^{-6} excess thyroid cancers per rad per year at risk was used (Ron et al., 1995) as the basis for determining the risk coefficients shown in Table 2.3. This risk is based on the pooled analysis by Ron et al. (1995) of epidemiological studies. This risk estimate is based on population studies of children exposed to external beam radiotherapy and Japanese survivors of the atomic bombings. The thyroid gland is especially vulnerable to the carcinogenic effects of radiation in children. Little risk is apparent in individuals exposed after age 20 years (Ron et al., 1995). Since the Alaska Native and white military personnel

TABLE 2.2. Risk of thyroid cancer from I^{131} exposure

Study Population	Age at Exposure (years)	Number of Irradiated Persons	Mean Dose (rad)	Absolute Risk[a] x10^{-6} per-person-year-rad)	Excess Relative Risk (per 100 rad)
Childhood exposures					
Swedish diagnostic I131	0-19	~2,000	160	0.2 (<0-0.9)1	0.5 (<0-2.6)
FDA diagnostic I131	0-20	3,503	~60	0.5 (<0-3.5)	3.1 (0-2.3)
Utah I131 fallout	0-9	1,962	~20	0.0 (0-5.6)	0.0 (0-3.7)
Marshall Islands	0-18	127	1,240	1.1 (0.4-2.3)	0.3 (0.1-0.7)
Adult Exposures					
Swedish diagnostic I131	>19	24,200	42	0 (<0-<0)	<0 (<0-<0)
German diagnostic I131	Adult	13,896	100	0.9 (0-2.6)	0.3 (0-1.4)
Marshall Islands	>18	126	25	1.3 (0.1-3.8)	0.5 (0.0-1.6)

[a]90 percent confidence limits in parentheses.

SOURCE: Shore, 1992.

TABLE 2.3. Thyroid cancer risk estimation in Alaska Natives and white military personnel

Population	Number of Subjects	Average Activity Administered (microcuries)	Dose Conversion Factor (rad/microcurie)[a]	Average Thyroid Dose (rad)	Risk Coefficient (lifetime excess cancer per million person-rad)[b]	Risk (probability of radiogenic thyroid cancer)
Wainwright males	39	27	1.3	35	2.8	1/10,000
Wainwright females	8	58	1.3	75	5.5	1/2,500
Point Lay males	7	16	1.3	21	2.8	1/17,000
Point Lay females	5	16	1.3	21	5.5	1/8,700
Anaktuvuk Pass males	13	74	2.9	215	2.8	1/1,700
Anaktuvuk Pass females	7	76	2.9	220	5.5	1/800
Fort Yukon males	6	55	1.3	72	2.8	1/5,000
Arctic Village males	5	68	3.8	258	2.8	1/1,400
Arctic Village females	6	70	3.8	266	5.5	1/700
Point Hope males	6	50	1.3	65	2.8	1/5,500
Total Natives	102					
White males (infantry-men and airmen)	19	58	1.3	74	2.8	1/4,800

[a]From calculations provided by Oak Ridge Institute for Science and Education, June 21, 1994 (see Appendix C).

[b]SOURCE: Shore (1992) and Ron et al. (1995). Risk coefficients are for adult males and females. Subjects are assumed to be at risk for 30 years (NCRP, 1985).

subjects in the AAL study were adults and were administered I^{131}, the thyroid risk coefficient of 4.4 x 10^{-6} was modified to account for these factors. The following factors were used to arrive at the risk coefficients listed in Table 2.3: a dose effectiveness reduction factor of one-fourth for I^{131} (Shore, 1992); a sex factor of two-thirds for males and four-thirds for females (NCRP, 1985; Ron et al., 1995); an age correction factor of one-eighth for adults (Shore, 1992); and 30 years at risk following exposure (NCRP, 1985).

RISK ESTIMATES FOR THE AAL STUDY

The AAL conducted four types of studies to measure (1) thyroid uptake, in which the percentage of I^{131} appearing in the thyroid gland was measured at various times after I^{131} administration, (2) iodine metabolism, in which urinary and salivary excretion rates and blood levels of I^{131} were measured at various times after I^{131} administration, (3) effects of exogenous potassium iodide treatment on thyroid uptake of I^{131}, and (4) iodine metabolism in white military personnel before and after a four-week cold exposure in the field.

Inspection of Table 2.3 reveals a wide range of thyroid doses and thyroid cancer risk estimates in the population groups participating in the AAL studies. Doses and risk estimates varied by more than a factor of 10. The lowest doses and risks were seen in Wainwright, Point Lay, Fort Yukon, and Point Hope Alaska Natives. Sixty-eight subjects from these communities participated in single studies and received a single administration of I^{131}. Twenty-two other Alaska Native subjects participated in two studies and received two separate I^{131} administrations. All 19 white servicemen participated in both the thyroid uptake and thyroid metabolic studies. Each serviceman received two doses of I^{131}. A group of 12 Anaktuvuk Pass Eskimos and Arctic Village Indians participated in three studies and received three separate I^{131} doses. Anaktuvuk Pass Eskimos and Arctic Village Indians had the highest thyroid doses and calculated thyroid cancer risks. For the subjects given multiple I^{131} administrations, all individual doses administered were added together and the total dose was used to estimate thyroid cancer risk. This assumes that there is no reduction in thyroid cancer risk due to protraction of the dose.

The AAL Technical Report (Rodahl and Bäng, 1957) identified two Wainwright subjects and one Arctic Village subject who were nursing children at the time of their participation. Since radioactive iodine can be passed on to children through the mother's milk, the childrens' thyroids may have been at risk. Thyroid activity in the nursing children was not measured directly in the AAL study. The AAL report indicated that 33 percent of the dose administered to the mother would be expected to appear in the mother's milk, and that uptake in the child's thyroid would be 33 percent of the ingested activity; thus the activity in children may be estimated to be approximately one-ninth of the activity administered to the mother. Based on these assumptions, an average thyroid activity in the nursing children of 12 rad was estimated. Because the resulting thyroid dose to these children is small, the thyroid cancer risk (about 1/2,000) is low.

Two young Alaska Natives participated in the AAL studies. One was a 16-year-old male Point Lay Inupiat and the other a female 17-year-old Arctic Village Athabaskan Indian. For the purposes of calculating risk, these subjects were considered adults since they were older than age 15 years (Ron et al., 1995) and thus physiologically closer to adults than children.

A 30-year-old Arctic Village Athabaskan Indian may have been pregnant at the time of her participation in the I^{131} studies, but the Committee has no way to verify this, (the possibility was raised by recent observers calculating backwards from the age of the participant's child, a relatively unreliable technique given the variability of pregnancy lengths). The embryo/fetus is particularly radiosensitive. The nature of the biological effects to be expected from an exposure during pregnancy depends on the magnitude of the dose to the embryo/fetus and when during gestation exposure occurs. If the participant was pregnant at the time she received a single dose of 50 microcuries in October 1955, the dose to the embryo is estimated to be 0.05 rad based on exposure to the embryo/fetus from accumulation of radioiodide in the mother's urinary bladder. The embryo/fetus would receive a negligible radiation dose from the mother's thyroid. Given the uncertainty, it is also unknown when this exposure might have occurred with respect to gestational age. However, because of the small dose, the risk of developmental abnormalities and other untoward pregnancy outcomes is small.

SIGNIFICANCE OF CALCULATED RISKS OF RADIATION-INDUCED THYROID CANCER

The last column of Table 2.3 provides lifetime estimates of the probability of thyroid cancer as a result of I^{131} administration. To put these risks into perspective, it is useful to consider the natural incidence of thyroid cancer in the population and the lifetime risk of thyroid cancer in the absence of radiation exposure. Thyroid cancer is a rare form of cancer (American Cancer Society, Inc., 1995). The American Cancer Society estimated there would be approximately 14,000 new cases of thyroid cancer in the United States in 1995. By comparison, 183,000 new cases of breast cancer, 170,000 new cases of lung cancer, and 138,000 new cases of colon-rectum cancers are estimated to occur during 1995 (American Cancer Society, Inc., 1995).

Assuming that thyroid cancer is uniformly distributed in a population of 260 million, the annual risk of thyroid cancer is about 5 cases per 100,000 population. Further assuming a 40-year period of risk (NCRP, 1985), the total lifetime background thyroid cancer risk would be 200 per 100,000 or 1 in 500.[4] The weighted average risk among the various populations

[4]New information from the Alaska Native Tumor Registry for the period 1969 - 1988 (Lanier et al., 1994) has demonstrated that Indian women have a higher incidence of thyroid cancer than the U.S. population as a whole, whereas the rates for Inuit men are lower, and the rates for Inuit women and Indian men are the same as for the U.S. population as a whole. Based on this report (ibid., p. 14), the annual average incidence of thyroid cancer for Inuit men is only 1 in 100,000 or a lifetime risk of 50 in 100,000 over 50 years, yielding a 1-in-2,000 risk. Indian women, based on this report, are much more susceptible to the disease, with an annual average incidence rate of 9.9 per 100,000 (ibid., p. 15) or a lifetime risk of 495 per 100,000 over 50 years, yielding a lifetime risk of approximately 1 in 200 (double the U.S. average). However, the total population of Alaska Natives is less than 100,000 and the Committee's review of the data showed that the data differences were not statistically significant in comparison with the overall U.S. population. Although the indicated increased risk of this disease in

(continued)

participating in the AAL study (Table 2.3) is about 1 in 3000, a risk six times lower than the background thyroid cancer risk. Thus, participation in the AAL study added a small and insignificant amount to the background thyroid cancer risk. The greatest risks (albeit small) of thyroid cancer were seen in populations given multiple I^{131} doses. In particular, the Anaktuvuk Pass females and Arctic Village females who received multiple doses have calculated risks of 1 in 800 and 1 in 700, respectively (Table 2.3). Thyroid cancer risk in these individuals is almost doubled because of the multiple I^{131} administrations. However, because thyroid cancer is rare to begin with, the additional radiological risk is not statistically significant. Radiation-induced thyroid cancers caused by the AAL study would not be expected in the Alaska Native or white military personnel experimental subjects. The Committee is unaware of any reports of thyroid cancer in the irradiated population. However, there has not been a systematic medical follow-up to determine whether any cases of thyroid cancer have appeared in the study subjects.

RADIATION GUIDELINES FOR I^{131} USAGE—THEN AND NOW

1957 Guidelines

At the time the AAL study was conducted in the mid-1950s, there were no formal guidelines concerning radiation exposure of research subjects. The AEC did approve the study, primarily based on radiological considerations; no radiation limits for diagnostic tests were in place, and I^{131} was the only radioactive material available to conduct the study. Setting aside the problems of informed consent, use of special populations, and methods of subject selection addressed elsewhere in this report, the study was scientifically reasonable by the standards of the time, assuming radiation exposure as the only consideration.

The prevailing view within the scientific community during the 1950s was that in order for radiation effects (particularly acute health effects such as reddening of the skin) to occur, the dose must exceed a threshold level. This threshold philosophy implied that radiation doses below the threshold were safe and did not cause harm. In interviews with Dr. Rodahl, it was clear that he accepted this philosophy then and still does now, and that he was convinced that the doses of I^{131} he used in the AAL thyroid uptake studies in the Inupiats, Athabascan Indians, and white military personnel were below the threshold and therefore perfectly safe. The researchers followed the threshold philosophy in administering multiple doses to some subjects. Although the cumulative effects of such multiple administrations were unknown at the time, the researchers believed that the effects of the first dose disappeared by the time of the administration of the second dose because the second dose was given several weeks after the first, allowing for complete decay of the I^{131} in the first administration. A review of medical literature of the period shows that the same general level of diagnostic dose was believed appropriate, though some researchers had begun using smaller amounts (Rall, 1956, 1957; Clark, 1956).

Athabascan women is a cause for concern, at this time the data are inadequate to warrant a revision in the current analysis.

In the 1950s, scientists believed that diagnostic doses of I^{131} were not associated with cancer risk. Small doses, as used in the AAL study, were believed to be safe because no evidence had emerged that small doses, as typically used in diagnostic studies, were associated with thyroid cancer. The Holm study (Holm et al., 1988) confirmed the relative safety of diagnostic doses of radioiodide. It should be noted here that even today there are no existing guidelines concerning the need for medical follow-up for persons given diagnostic doses of I^{131}. However, the use of pregnant a possibly pregnant woman would have been considered inappropriate based on generally accepted practice in the 1950s (Clark, 1956), and lactating women would have been discouraged from breast feeding until the I^{131} cleared their system.

Giving I^{131} to children was being seriously questioned at the time of the AAL research (Clark, 1956; Rall, 1957) but was not yet recommended against as a standard. The possibility that thyroid cancer could result from radiation exposure was first documented in 1950, but those cases resulted from large doses used for therapeutic purposes. In 1955, Simpson and colleagues conducted one of the earliest studies linking thyroid cancer in children with external radiation treatment for thymic enlargement (Simpson et al., 1955). At about the same time that studies of thyroid neoplasms subsequent to x-ray therapy for benign disease were going on, the possibility that incorporation of radioiodide in the thyroid could also be carcinogenic was also explored. Studies of the Marshall Islands inhabitants who were exposed to radioactive fallout (including radioiodide) from a thermonuclear test in 1954 and patients given high doses of radioiodide for thyroid therapy (e.g., hyperthyroidism) suggested that radioiodide incorporation could also be carcinogenic. The Marshallese were among the most intensely studied of any group subjected to thyroid irradiation and provided the first suggestion of a link between radioiodide incorporation and thyroid cancer (Shore et al., 1986).

Appendix D contains the Air Force research protocols for the use of human volunteers in experimental research as set forth in a 1953 memo.

Current Guidelines

This Committee was asked to address the risks to the AAL study participants under today's scientific standards. The Committee believes the AAL experiment would not be allowed by an Institutional Review Board (IRB) today because of the radiation dose to the thyroid from the I^{131} to be used. The dose estimates for every group of subjects listed in Table 2.3 exceed the current recommended radiation dosage limits approved by the FDA for research subjects (5 rad for a single dose, 15 rad as a cumulative dose for a number of studies conducted within one year).

While there were various problems with the subject selection process, discussed elsewhere in this report, from a scientific perspective minors, lactating women, and potentially pregnant women would probably be excluded from this type of study today. The use of research subjects with obvious or modest thyroid enlargement would also have been considered a questionable practice. In addition, while the final report of the study stated that the participants were healthy, "normal" individuals, the Committee found that the physicians took the people who came to them and then performed (but not even in all cases) a rudimentary physical examination consisting of measuring height, weight, oral temperature, pulse, and blood pressure.

The study design stated: "Care was taken to exclude individuals who had taken medication or x-ray contrasts which are known to affect the iodine uptake" (Rodahl and Bäng, 1957, p. 6). No Native participants could recall being asked about their physical condition or if they were taking medicines. The Committee did not speak to enough military participants to draw conclusions about their selection. The Committee believes the doctors probably had access to the military participants' medical records to determine their medical history, but such records were unlikely to exist for the Native subjects. (See Box 2.3 for additional discussion of use of I^{131} in special populations.)

THE EVOLUTION OF OUR UNDERSTANDING OF RADIATION HEALTH EFFECTS

It was during the late 1950s that the no-threshold philosophy began to emerge and the importance of chronic health effects such as cancer was recognized. Today, radiation protection and radiological health principles are firmly grounded in the no-threshold philosophy: any dose of radiation is potentially harmful. The specific probability of harm is dependent upon the radiation dose received. Unfortunately, this no-threshold philosophy is often interpreted to mean that no radiation dose is "safe." As exemplified by this study, this interpretation of the concept of threshold is incorrect. Although a particular risk may be attributed to a given radiation dose, the probability of thyroid cancer is so small as not to be measurable in the population. The question of what is a "safe" dose is trans-scientific and requires consideration of social and economic factors in addition to dose-response data. Recently some members of the radiation science community have begun to question the assumptions giving rise to the linear no-threshold philosophy because there are experimental data that can be interpreted as supporting alternative hypotheses. This question will continue to be considered and debated in the future.

In 1950, the International Commission on Radiological Protection (ICRP) established an occupational exposure standard of 0.3 roentgen per week. This dose was based on the concept of tolerance doses related to the development of erythema, or skin reddening. ICRP exposure limits for radiation workers were based on the observation that many workers who had been in contact with radiation for a number of years at these levels had suffered no apparent radiation injury (Hendee, 1993). During the 1950s, information about the importance of cancer as a delayed effect of radiation began to emerge from studies of the Japanese survivors of the atomic bombings at Hiroshima and Nagasaki in 1945. What emerged from the studies of the Japanese survivors and patients who had received radiation for the treatment of various medical conditions was the belief that any dose of radiation may be harmful and that the major health effect of exposure was cancer. As a consequence, occupational exposure standards were reduced to 5 roentgens per year in 1956. These limits were applicable only to individuals occupationally exposed to radiation. Human subjects exposed to radiation as part of a medical diagnostic or therapeutic procedure or who participated as volunteers in medical research studies were not subject to these limitations.

Box 2.3
STANDARDS FOR USE OF I^{131} IN SPECIAL POPULATIONS

Mid-1950s Standards

An extensive search of medical literature of the time indicates that radiation injury from the use of even diagnostic amounts of 1^{131} was a topic of concern during the 1950s. A key article in the American Journal of Medicine stated "...it is easy to forget that a IO tracer...represents the amount of radiation that would normally be permitted as the 'maximum permissible safe dose' throughout the course of an entire year... Radioiodine tracers to children especially should be viewed with considerable caution" (Rall, 1956, p.729). We were unable to locate references of the time that restricted the use of the tracer in lactating women, though it was apparently a developing practice to do so. The concern over irradiation and potential genetic damage to the fetus was an important topic of the *Conference on Radioiodine* (Clark, 1956) which was conducted under the auspices of the AEC in Chicago in 1956. At that conference hospital administrators stated they either did not allow the use of radioiodine with pregnant women, or advised abortions if a woman was pregnant at the time of radioiodine testing.

Current Standards

Pregnancy and lactation are considered absolute disqualifications for the use of I^{131} in treatment of hyperthyroidism (Edmonds and Smith, 1986), although the same is not true for microcurie quantities for diagnostic use. (Therapeutic levels of I^{131} are much higher than diagnostic levels.) However, most medical centers have avoided use of I^{131} in pregnant and lactating women even for diagnostic purposes. Radioiodide is concentrated in the breast tissue, especially during lactation (Baeumler, 1986; Dydeck and Blue, 1988; Hedrick et al., 1987; Lawes, 1992; Romney et al.; 1989). Therefore, in cases where a lactating mother has had to have doses of radioiodide administered to her, it has been recommended that the mother stop breastfeeding for up to 14 days (Lawes, 1992; Romney et al., 1989). Some suggest that no radioiodide studies should be done in women who wish to continue breastfeeding (Dydeck and Blue, 1988).

Also, radioiodide is transported freely across the placenta into the fetus. Administration of I^{131} doses during pregnancy can adversely affect and even ablate the fetal thyroid after 10 weeks of gestation, when the fetus. Administration of I^{131} doses during pregnancy can adversely affect and even ablate the fetal thyroid after 10 weeks of gestation, when the fetal thyroid becomes capable of accumulating iodine (Green et al., 1971; Shepard, 1967). One case has been described in a woman who was administered 1^{131} at therapeutic levels inadvertently during the first week of pregnancy and a high concentration of radioactivity was observed in the pregnant uterus (Cox et al., 1990). Exposure to radiation was considered to be a factor involved in the occurrence of abortion, in this case, at 8 weeks of gestation (Romney et al, 1989).

For men, literature review suggests that 400 rads or more of radiation dose is required to threaten fertility. This degree of radiation exposure occurs only after therapeutic doses of I^{131}, not at the levels recorded in the AAL experiments. Fertility has been observed to remain normal even after therapeutic dosage (Edmonds and Smith, 1986) even though it is known to potentially cause significant oocyte loss from the gonads (Baker, 1971). A high dose of I^{131} ($\sim$350 mCi), as needed for treatment of thyroid cancer, was reported to have been followed by occurrence of testicular failure in one case (Ahmed and Shalet, 1985).

On July 25, 1975, the FDA established for the first time limits on radiation to adult human volunteer research subjects (Department of Health, Education, and Welfare, 1975). The amount of radioactive material administered to human research subjects during the course of a research project intended to obtain basic information regarding the metabolism (e.g., kinetics, distribution, and localization) of a radioactively labeled drug should be the smallest radiation dose that can be administered without jeopardizing the benefits to be obtained from the study. The limit for the thyroid gland is 5 rad as a single dose and 15 rad as a cumulative dose from a number of studies conducted within one year. For children (research subjects under 18 years of age), limits are 10 percent of adult limits. Thyroid doses below these levels are generally recognized as safe (Department of Health and Human Services, 1990; Mossman, 1992).

3

The Ethics of Human Subjects Research

This Committee was asked to determine whether the AAL thyroid function study "was conducted in accordance with generally accepted guidelines for use of human participants in medical experimentation," and whether participants had been notified about possible risks. This aspect of the Committee's charge proved to be particularly challenging. Determining whether the I^{131} research was conducted according to generally accepted guidelines of the 1950s means first determining what research ethics guidelines should be applied to judge human subjects research that took place at that time in history, and second, applying those general guidelines to the unique circumstances of the I^{131} study. To do this with any accuracy, it is necessary to consider the history and evolution of medical ethics and the ethics of research with human subjects. We must understand where we were at the time the research was conducted in comparison to where we are today, and face the problems inherent in second-guessing the past.

To determine whether the AAL thyroid function research followed "generally accepted guidelines," the Committee had to consider both principles and practices. If only the principles of research ethics were considered, then virtually all research would be unethical, as principles can only rarely be employed in a manner that raises no questions at all. Thus it seems reasonable to conclude that considering only principles would hold researchers to an unfairly high standard. If, however, only the actual practices of the time were considered, a different unfairness would result. Requiring researchers only to do what other researchers do would render principles meaningless and change impossible. Thus, considering only practices would justify all research, and such a standard would be unfair to subjects.

The challenge, then, when assessing research with human subjects—whether the research is in the past or the present—is to examine both principles and practices to determine and apply "generally accepted guidelines." This is what the Committee sought to do. In the process, we realized that the evolution of principles and practices of human subjects research has been more gradual than swift, and composed of small and spreading changes rather than of dramatic paradigm shifts. As important as our examination of the I^{131} study is, it is even more important to recognize that education about the ethics of human subjects research is a continuing necessity today.

BACKGROUND

To understand the concept of informed consent and the legal/moral underpinnings of human subject experimentation as related to the AAL study, it is necessary to understand how these underpinnings developed. Medical treatment and human subjects research share two ethical mandates: to avoid harm and to respect the patient's self-determination (autonomy). Thus, research must be designed to produce meaningful knowledge with minimal risk. In addition, subjects must make free and informed decisions about whether to participate. These issues are discussed at length in *A History and Theory of Informed Consent* (Faden and Beauchamp, with King, 1986), which is drawn upon heavily here.

The ethical mandates underlying the conduct of research using human subjects are derived from several sources, including:

- the ancient and traditional duty of physicians to benefit their patients, or at least do them no harm;[1]
- the Kantian philosophical view of human beings as "ends in themselves," never to be used merely as means to ends, or for the advantage of others;[2]
- the political and legal concept of autonomy or self-determination that requires consent to any bodily intrusion;[3] and
- the requirements of good scientific method in designing and conducting experiments, including: minimization of risk.[4]

These ethical concepts are often translated into three basic principles that provide a framework for the moral conduct of human subjects research:

- The principle of "autonomy," or personal self-governance, "by adequate understanding while remaining free from controlling interferences by others and from personal limitations that prevent choice" (Faden et al., 1986, p. 8). In order for a research subject to make an autonomous choice, the autonomy of the subject must be respected, which includes providing sufficient information for the subject to make an autonomous and informed decision.

---

[1]According to Faden et al. (1986, p. 10), "[A] celebrated principle in the history of medical codes of ethics is the maxim *primum non nocere*—'above all, do no harm.' Recent scholarship has shown that in the Hippocratic writings the more precise formulation is 'help, or at least do no harm,' thus demanding the provision of benefit beyond mere avoidance of harm."

[2]According to Faden et al. (1986, p. 8). "As expressed in Kantian philosophy, autonomous persons are ends in themselves, determining their own destiny, and are not to be treated merely as means to the ends of others. Thus, the burden of moral justification rests on those who would restrict or prevent a person's exercise of autonomy."

[3]Faden et al., 1986, p. 1, p. 7-9, 26-30.

[4]Federal Common Rule §111(a)(1)-(2). See Appendix E.

- The principle of "beneficence," which is concerned with the intent and capacity of science and medicine to avoid harm and provide benefit; in the case of research, this requires careful weighing of potential harms against potential benefits.
- The principle of "justice," or treatment according to what is fair, due, or owed, which includes avoiding unfairly burdening subjects or communities of subjects in relation to benefits.

For every research project involving human subjects, two basic inquiries are necessary. One inquiry must examine the necessity of the research, the expected results, the risk-benefit balance, and minimization of risks. The other inquiry must examine the fairness of subject selection, adequacy of information given to prospective subjects, and voluntariness of subjects' consent to participation. In very general terms, the first inquiry addresses the research's potential for *harming* subjects and the second addresses the research's potential for *wronging* them. Also speaking very generally, the first inquiry is beneficence based, whereas the second is concerned with autonomy and justice.

Although these two sets of research components are considerably interdependent, it is nonetheless possible to commit harms without wrongs—for example, an unexpected injury can occur in an apparently risk-free experiment before it can be halted. It is also possible to commit wrongs without harms, such as when a minimally risky experiment causes no injuries to subjects, but where the subject selection process or the informed consent process is defective.

Harming without wronging may be easier to grasp than wronging without harming. In biomedical settings, the focus on avoiding physical harms can tend to overwhelm other considerations. However, especially in research, according participants full respect and dignity and treating them with fairness and justice is of great importance, independent of physical harm. In some contexts, treating study participants with less than full respect may be more damaging than physical harm, and the effects may be longer lasting. The Committee, therefore, gave equal consideration to the potential for harms and for wrongs.

Evolution of Ethical Standards

Although use of human subjects in medical experiments is an ancient practice, use of human subjects in scientifically rigorous research only became established in the United States shortly before the outbreak of World War II. Before the postwar promulgation of the Nuremberg Code, some attention was paid to the ethical issues raised by medical research with human subjects (Faden et al., 1986, pp. 151-153). Early moral and legal concerns about human subjects research focused more on controlling research risks than on enabling autonomous choice by research subjects. It took the Nuremberg Trials, however, to bring human subjects research to public consciousness and to launch the development and implementation of ethical standards for such research.

Although the evolution of standards for the use of human subjects in research does not start with the Nuremberg Code, the Code was a watershed event in bioethics that helped determine the development of contemporary standards. It is essential to understand that the Nuremberg Code was intended, and was understood at the time, to be an expression of existing and universal moral principles governing research with human subjects (Advisory Committee on Human Radiation Experiments, 1995). Therefore, despite the fact that the Code had not existed as such at the time the Nazi experiments were conducted, the Nuremberg Military Tribunal could apply it to condemn the experiments.

The Nuremberg Code

The Nuremberg Military Tribunal was formed to examine the legal and ethical aspects of human experiments carried out by Nazi physicians before and during World War II. As a result of its investigation, on August 19, 1947, the Tribunal issued a 10-point list of principles delimiting morally and legally permissible human experimentation (see Box 3.1). The Code focuses primarily, but not exclusively, on the subject's free and informed consent. It requires subjects to have decision-making capacity and to be able to consent freely, without "the intervention of any element of force, fraud, deceit, duress, over-reaching, or other ulterior form of constraint or coercion." Subjects must be given information about the nature, methods, duration, and purpose of the experiment, and about all reasonably anticipated risks, harms, inconveniences, and health effects, in order to enable them to choose with understanding about their participation. Subjects must be able to withdraw from participation at any time. In addition, the Code prescribes that research design should ensure that the experiment is capable of yielding worthwhile results, that the degree of risk does not exceed the value of the results, that all risks to human research subjects are minimized, and that research is conducted according to appropriately high scientific standards.

After the Code

The Code was well promulgated and widely discussed, beginning in the late 1940s and early 1950s. In 1946, while the Nuremberg Trials were going on, the Judicial Council of the American Medical Association (AMA) set out three requirements for human subjects research: (1) voluntary consent of the subject, (2) prior animal experimentation to determine risk, and (3) proper medical management of the experiment.[5] The February 27, 1953, issue of *Science* led with a set of short papers from a 1951 symposium under the title "The Problem of

[5]These requirements constituted the Judicial Council's application of the AMA's Principles of Medical Ethics to human experimentation. The requirements were issued in response to Dr. Andrew C. Ivy's report on the Nuremberg Trials. Dr. Ivy served as the AMA's official consultant to the Nuremberg prosecutors and played a significant role in the development of the Nuremberg Code (Advisory Committee on Human Radiation Experiments, 1995). They were published in the *Journal of the American Medical Association* (JAMA) on December 28, 1946, and reiterated in 1958 when JAMA published the AMA's 1957 Principles of Medical Ethics and Opinions and Reports of the Judicial Council.

Box 3.1
NUREMBERG CODE
1946

The Nuremberg Military Tribunal's decision in the case of the United States v. Karl Brandt et al. includes what is now called the Nuremberg Code, a 10-point statement delimiting permissible medical experimentation using human subjects. According to this statement, humane experimentation is justified only if its results benefit society and it is carried out in accord with basic principles that "satisfy moral, ethical, and legal concepts."

1. The voluntary consent of the human subject is absolutely essential. This means that the person involved should have legal capacity to give consent; should be so situated as to be able to exercise free power of choice, without the intervention of any element of force, fraud, deceit, duress, over-reaching, or other ulterior form of constraint or coercion; and should have sufficient knowledge and comprehension of the elements of the subject matter involved as to enable him to make an understanding and enlightened decision. This latter element requires that before the acceptance of an affirmative decision by the experimental subject there should be made known to him the nature, duration, and purpose of the experiment; the method and means by which it is to be conducted; all inconveniences and hazards reasonably to be expected; and the effects upon his health or person which may possibly come from his participation in the experiment.

   The duty and responsibility for ascertaining the quality of the consent rests upon each individual who initiates, directs or engages in the experiment. It is a personal duty and responsibility which may not be delegated to another with impunity.

2. The experiment should be such as to yield fruitful results for the good of society, unprocurable by other methods or means of study, and not random and unnecessary in nature.

3. The experiment should be so designed and based on the results of animal experimentation and a knowledge of the natural history of the disease or other problem under study that the anticipated results will justify the performance of the experiment.

4. The experiment should be so conducted as to avoid all unnecessary physical and mental suffering and injury.

5. No experiment should be conducted where there is an *a priori* reason to believe that death or disabling injury will occur, except, perhaps, in those experiments where the experimental physicians also serve as subjects.

6. The degree of risk to be taken should never exceed that determined by the humanitarian importance of the problem to be solved by the experiment.

7. Proper preparations should be made and adequate facilities provided to protect the experimental subject against even remote possibilities of injury, disability, or death.

8. The experiment should be conducted only by scientifically qualified persons. The highest degree of skill and care should be required through all stages of the experiment of those who conduct or engage in the experiment.

9. During the course of the experiment the human subject should be at liberty to bring the experiment to an end if he has reached the physical or mental state where continuation of the experiment seems to him to be impossible.

10. During the course of the experiment the scientist in charge must be prepared to terminate the experiment at any stage, if he has probable cause to believe, in the exercise of the good faith, superior skill and careful judgment required of him that a continuation of the experiment is likely to result in injury, disability, or death to the experimental subject.

["Permissible Medical Experiments." *Trials of War Criminals before the Nuremberg Military Tribunals under Control Council Law No. 10: Nuremberg October 1946-April 1949.* Washington: U.S. Government Printing Office (n.d.), vol. 2, pp. 181-182.]

Experimentation on Human Beings." These papers included an extensive discussion of the Nuremberg Code, as well as a discussion of rights of military personnel regarding medical treatment and research. The Code's general principles were adopted specifically by both the AMA and the Department of Defense.

But in the postwar period, dissemination of and implementation of the Nuremberg Code was, to say the least, uneven. In 1953, the Department of Defense formally adopted the Code in guidelines addressing the use of human subjects for research related to atomic, biological, and chemical warfare, but the document was classified Top Secret because of government sensitivity about these military issues (see Appendix D). This is the earliest instance in which a federal agency that sponsored human experiments adopted the Nuremberg Code, although related discussions were ongoing in other agencies in the same era (Advisory Committee on Human Radiation Effects, 1995). (The policy remained classified until 1975.) The Atomic Energy Commission discussed informed consent issues in a series of letters dating back to 1947 and developed subject consent requirements for the use of radioisotopes it supplied to medical researchers, but did not systematically promulgate or enforce them (Advisory Committee on Human Radiation Experiments, 1995). Many medical researchers apparently believed that because the Nazi experiments were so obviously flawed, both ethically and scientifically, the Nuremberg Code was intended only to apply to ill-intentioned human subjects research—"It was a good code for barbarians but an unnecessary code for ordinary physicians" (Advisory Committee on Human Radiation Experiments, 1995).

At the same time, worldwide attention was being paid to the language of the Code itself, and new guidelines were being drafted to provide better guidance for physicians and researchers than the Code was thought to offer. Professional societies and associations, world bodies, and governmental entities all began to develop guidelines. One set of guidelines became especially influential. This was the Declaration of Helsinki, first drafted in 1961 by the World Medical Association. Like the Nuremberg Code, the Declaration of Helsinki focused on consent as a central requirement (Faden et al., 1986, pp. 156-157).

In 1953, the National Institutes of Health's new Clinical Center developed and implemented a rigorous in-house policy requiring informed consent and peer review of risk-bearing research (Faden et al., 1986, pp. 201-202). And in 1959, Henry Beecher published the first of his influential exposes of research practices with human subjects, the monograph "Experimentation in Man" (Faden et al., 1986, p. 157). Thus, in the early to mid-1950s the principles governing research with human subjects were firmly in place, but their implementation in practice was incomplete and even confused. Clear moves toward systematic reform of research practice did not come until the 1960s and 1970s.

The Evolution of Modern Federal Regulations

In the United States, modern guidelines for human subjects research began to take final form in 1974. In 1971 the U.S. Department of Health, Education, and Welfare (DHEW) issued guidelines for the conduct of social and behavioral research that in 1974 were codified into federal regulations governing human subjects research in institutions receiving DHEW funds. Also, legislation (P.L. 93-348, National Research Act) created the National Commission for the Protection of Human Subjects of Biomedical and Behavioral Research, which in 1979 authored what became know as the *Belmont Report* (National Commission for the Protection of Human Subjects of Biomedical and Behavioral Research, 1979).

The *Belmont Report* sets forth the same three basic ethical principles governing human subjects research that were outlined early in this chapter and that underlie the Nuremberg Code: respect for persons, beneficence, and justice. The report points out that respect for persons demands the informed and voluntary consent of the subjects; beneficence demands an assessment of the relative probability and magnitude of risks to subjects against benefits expected from the results; and justice demands equitable selection of subjects and equitable distribution of the research's benefits and burdens between research subjects and the population at large.

The 1974 DHEW regulations[6] grew into a federal policy for the protection of human

---

[6]These regulations establish a system of Institutional Review Boards (IRBs) charged to review and approve of human subjects research, and set forth approval criteria. These criteria include the following: that risks to subjects be minimized; that risks to subjects be reasonable in relation to anticipated benefits (continued)

subjects known as the "Common Rule." It is applicable to all human subjects research taking place under the auspices of any federal department or agency, including the Department of Defense. The federal Common Rule was published in the *Federal Register* on June 18, 1991 (56 F.R. 28003-28032). (The Department of Defense's version of the Common Rule appears in its regulations, Code of Federal Regulations, volume 32, part 219.) The Common Rule provides much more detailed and specific guidance to human subjects researchers and their institutions than was available to the AAL physicians. The regulations develop the principles of the Nuremberg Code and provide guidelines for their implementation, which the Code lacked. In 1993, there was much more information available to assist researchers and their institutions in implementing appropriate protection for human subjects than was available to the AAL researchers in the 1950s, but the basic underlying principles were the same then as now.

The Nuremberg Code as Applied to AAL Research in 1955-1957

During the mid-1950s, the application of ethical standards for the conduct of human subjects research was uneven in the federal government and guidelines for implementation were inconsistent. Many researchers, policymakers, and governmental and military personnel may have mistakenly considered the Nuremberg Code to apply only to overtly hazardous research or experiments lacking scientifically reasonable goals. As related earlier, Dr. Rodahl, when asked directly over the telephone during the public hearing in Fairbanks if he had any guidelines similar to the Nuremberg Code for obtaining informed consent, told the Committee that he did not, but that he had approvals from his supervisors and if there was anything wrong with the research it would not have been approved, and he would not have been allowed to publish his study results.

However, the Nuremberg Code itself, in its own language and as set forth in the medical and scientific literature of the times, is unqualified in its application to "the human subject." The Code was developed to be an expression of universal moral principles governing research with human subjects—it was considered to represent current research ethics, acceptable and accepted at the time rather than new terrain. Therefore, in the Committee's, opinion the standards outlined in the Code did apply to human subjects research in the 1950s, including research conducted under military auspices. Regardless of misconceptions about the applicability or scope of the Nuremberg Code, and regardless of the failure of the Air Force to implement

---

to the subjects or from the resulting knowledge; that subject selection be equitable; and that the subjects' informed consent be obtained. Informed consent requirements are set out, including a list of required elements of disclosure. The regulations are supplemented by publications from the National Institutes of Health Office for Protection from Research Risks, which include interpretive guidelines for use by IRBs in the conduct of their work. These guidelines address a variety of questions in detail, including subject selection, research with special populations, and refinements of informed consent.

Current AMA (1992) policy on clinical investigation similarly contains requirements for informed consent and "rigorous scientific standards" for research, but with considerably less detail than the federal regulations.

or disseminate its classified Code-based policy, in the Committee's opinion, the standards of the Code applied to the ethical conduct of the AAL research physicians and scientists, and to their superiors, at the time the research took place.

As should be clear from the Committee's description of the AAL research and the process of our inquiry, the I^{131} research was far from egregious. Subjects, both Alaska Natives and Air Force personnel, were given some information and an attempt at obtaining informed consent was made. The radioactive tracer was administered in doses believed to be harmless, and the purpose of the research was not to determine the effects of radiation in the human body[7]. Thus, by comparison with some of the Cold War research examined by the Advisory Committee on Human Radiation Experiments[8] (e.g., plutonium injections and active deception of patient-subjects), it might seem overly scrupulous to be concerned with a close analysis of the I^{131} study.

Nonetheless, there are three reasons for careful examination of the I^{131} study according to the terms of the Nuremberg Code. First, the subjects were normal healthy volunteers and there was no indication that the researchers expected the study to result in improvement of the subjects' health.[9] Second, radioisotopes are potentially harmful substances, and thus the I^{131}

---

[7]It should be noted that the AAL thyroid function study was only one of many studies using I^{131} conducted from the 1940s to 1960s. The report *Human Radiation Experiments Associated with the U.S. Department of Energy and Its Predecessors* (1995), prepared as part of DOE's effort to document human radiation experiments, includes at least 60 studies involving I^{131} in diagnostic, therapeutic, and research settings.

[8]The Advisory Committee on Human Radiation Experiments was appointed by President Clinton in 1994 to conduct an extensive inquiry into the history of government-sponsored human radiation experiments and intentional environmental releases of radiation that occurred between 1944 and 1974. The 1,000-page final report found that the government sponsored approximately 4,000 radiation experiments involving tens of thousands of individuals, many of whom had no knowledge that they were participating in radiation research. The committee recommended that the government apologize to people who received no direct medical benefit from participation in research and provide financial compensation for those who suffered physical injury as a result of participation. The committee recommended that subjects in three particular studies (about 30 people) receive financial compensation for being subjected to dangerous experiments without their knowledge, and gave guidelines that might lead to compensation to others. The committee rejected proposals that the government notify all known participants or their families or provide wide-scale medical follow-up. In October 1995, President Clinton accepted the committee's report and recommendations, and issued an apology to all those who were used as subjects in government-sponsored radiation experiments, and their families. He also established a new body, the National Bioethics Advisory Committee, to develop new policies to guide research in human biology and review ongoing government research.

[9]The Advisory Committee on Human Radiation Experiments (1995), in discussing the use of healthy subjects in medical research, noted: "The use of *patients* in medical research appeared in a different historical context from that of *healthy subjects*. From the perspective of the medical profession, the age-old tradition of the doctor-patient relationship provided a justification for research with the potential to benefit patients, but not, of course, for healthy subjects who were not under medical care."

research was not without theoretical risk. And third, the overwhelming majority of the subjects were Alaska Natives, who may have represented a population of convenience and whose language and cultural differences from the physician-researchers certainly affected the consent process and their role as subjects. It is to elucidate these reasons, and to explore their importance, that we embark upon an application of the terms of the Nuremberg Code to the I^{131} research.

Several aspects of the informed consent[10] relationship between researchers and subjects must be examined according to the terms of the Nuremberg Code: (1) subjects' legal capacity and freedom to choose under the circumstances; (2) researchers' disclosure of information about the experiment, and subjects' knowledge and comprehension of this information; and (3) the duty and responsibility of researchers to obtain consent.[11] Although the Code is general and offers no guidance as to how the requirements should be fulfilled, its language is explicit. The following discussion looks at the principal terms of the Code as relevant to the AAL research.

1. *"[T]he person involved should have legal capacity to give consent, [and] should be so situated as to be able to exercise free power of choice, without the intervention of any element of force, fraud, deceit, duress, over-reaching, or other ulterior form of constraint or coercion."*

The conduct of the AAL thyroid function study raises a variety of concerns related to consent and free power of choice. For instance, although the researchers excluded young children from the study, at least seven of the Alaska Native subjects were under 21, the age of majority in the Territory of Alaska at the time (Compiled Laws of Alaska, §20-1-1, 1949). Minors, unless "emancipated," lack the legal capacity to give consent. Minors become legally emancipated by participating in certain adult activities such as marriage or military service. Some of these minor subjects of the AAL research were emancipated by marriage; others were not. Parental consent was not sought for any minors, to the Committee's knowledge. The researchers apparently instructed elders and interpreters that they needed "adult or near-adult" volunteers, and then simply recorded the ages of subjects brought to them. The inclusion of minor subjects in research without parental consent is a violation of ethics and law, both for the period of the experiments and in 1993. There were three minor subjects among the U.S. Air Force participants; however, since military service is an emancipating condition, there was no legal conflict in those cases.

In addition, the Committee has significant concerns about whether the Alaska Native subjects were "so situated as to be able to exercise free power of choice" under the circumstances. These concerns relate to the use of village authority figures to identify

---

[10]Although the term "informed consent" did not come into use until 1957 (Faden et al., 1986, p. 125), the elements that define consent in the Nuremberg Code correspond well to the Committee's contemporary concept of "informed consent," and so that term is used here for convenience.

[11]There are many modern lists of elements of informed consent to research that are equivalent. For example, a moral definition is: An intentional authorization given to a researcher by a subject who has substantial understanding, and is not substantially controlled by others (Faden et al., 1986, p. 278). A policy definition is: The competent subject's voluntary consent after disclosure that the intervention involves research and about its nature, consequences, risks, benefits, and alternatives (Federal Common Rule, §116).

participants, language barriers that led to confusion over whether subjects were participating in research or medical treatment, and cultural differences that meant the Alaska Native subjects were hesitant to refuse participation. These issues are further elaborated in the justice section at the end of this chapter.

2. *"[T]he person involved should have sufficient knowledge and comprehension of the elements of the subject matter involved as to enable him to make an understanding and enlightened decision. This . . . requires that before the acceptance of an affirmative decision by the experimental subject there should be made known to him the nature, duration, and purpose of the experiment; the method and means by which it is to be conducted; all inconveniences and hazards reasonably to be expected; and the effects upon his health or person which may possibly come from his participation in the experiment."*

Dr. Robert Elsner (one of the three former AAL researchers who attended the public hearing and the former director of the AAL analytical laboratory) and Dr. Rodahl told the Committee that, to their recollection, fairly complete information was conveyed to both Alaskan Native and military subjects. Dr. Rodahl said he informed the village elders about the purpose of the study and what they wanted the volunteers to do, and that the elders were present as interpreters while the procedures were being done. Dr. Elsner, who worked primarily with military subjects, stated his belief that as much information as possible should be provided to all subjects. However, the interpreters' command of English was variable and in most cases insufficient to adequately explain the research. In addition, the interpreters may not have understood that they were recruiting subjects for research, rather than selecting villagers for medical treatment. Finally, neither military subjects nor Alaska Native subjects were informed that the I^{131} capsules they swallowed were radioactive; this stemmed from the researchers' conviction that I^{131} was harmless and therefore not an important element of disclosure. The lack of terms for "radiation" was an additional barrier to information conveyance in Native languages. The Committee believes that it might reasonably be asserted that experimental administration of a substance known to be dangerous at higher doses should have been accompanied, even in the 1950s, with disclosure about the possibility of unknown future risks. Such caution would be required today for research.

Based on the hearing testimony and field interviews at Native villages, the Committee firmly believes that most of the Native participants were not sufficiently briefed on the nature of the research and that some may not have understood that the study was research rather than treatment intended for their benefit (see Appendix B). The AAL research took place within a general context that included multiple visits by a variety of authority figures to Alaska Natives, for study, treatment, and other reasons. In addition, the basic framework of the AAL field investigations into acclimatization tended to blur distinctions between therapeutic and nontherapeutic medical actions, between anthropological and medical research, between essentially risk-free and potentially risk-bearing research, between research and the provision of basic medical care, and between researcher and physician. For example, the AAL physicians with whom the Committee spoke during its meeting in Alaska stated that they ministered to sick and injured villagers as soon as they came to a village because there was so little available care. After giving such care, they then went on to conduct their research studies.

This combination of research, treatment, and other interventions created confusion about whether a particular intervention was research or treatment, and bolstered the authority of researchers to obtain participation. If subjects of research are permitted to infer that they are receiving treatment intended for their benefit rather than participating in nontherapeutic research, there is a fundamental flaw in the foundation upon which their understanding and consent are based.

In the Committee's view, the researchers' task, even during the 1950s, was to enable subjects to understand; researchers may not merely assume their subjects' understanding. The AAL researchers should have recognized the possibility of gaps in subjects' understanding and should have determined whether it was possible to eliminate or reduce those gaps. If the Alaska Native subjects could not "make an understanding and enlightened decision" under the circumstances, the research should not have taken place. It must be emphasized that researchers, are not—indeed cannot be—morally required to *ensure* their subjects' understanding (see Faden et al., 1986; Beauchamp and Childress, 1994). The Committee's criticism of the AAL researchers is based on their lack of efforts to attend to, or reduce, linguistic and cultural barriers—*not* on their failure to succeed.

In the Committee's opinion, what the researchers obtained was closer to mere cooperation than informed consent. Given the cultural attitudes of the time, the AAL researchers may genuinely, though mistakenly, have believed that the subjects' acquiescence was sufficient under the circumstances. Perceptions about justice, cultural bias, and social sensitivity have developed considerably since the 1950s, and are discussed later in this chapter.

Most subjects who spoke with the Committee and to the team that went to the Native villages asserted that they believed they were receiving treatment, and if they had understood that the procedures they underwent were not for their benefit, they would not have participated. The Committee is unable to determine the extent to which the subjects' recollections, as expressed in their testimony, is colored by increased understanding or fear of radiation. However, the Alaska Native belief system lends credence to their assertion. Belief systems at the time were a mixture of traditional, supernatural, and Christian spiritual controls. Man was seen as an integral part of his earthly surroundings. Although a doctor or dentist could treat the physical ailments of a human, a shaman provided the way to communicate with spirit beings for treating the forces of nature and ensuring protection from harm and disease (Marsh, 1954). These belief systems affected how the Alaska Natives perceived the causes of illness. The traditional concept of illness was based on soul-loss, intrusion of a foreign object by sorcery (Chance, 1961), or breaking a taboo (Burch in Smithsonian Institution, 1984, p. 344). Thus, for Alaska Natives the lack of direct benefit to subjects from the research and the possibility of unknown risks would be associated with the likely occurrence of future harms to subjects or their families. These harms would appear to the researchers as utterly unrelated to the research, but to Alaska Natives could stem naturally from participation.

Although the Committee believes that the researchers did offer individuals an opportunity to decline to participate, the offer came after the subjects had been recruited and were about to be tested, making refusal awkward. The Nuremberg Code requires disclosure "before the acceptance of an affirmative decision" by the subject.[12] Moreover, to decline to participate in research is very different from declining what is believed to be beneficial treatment. A subject who misunderstood the nature of the intervention, believing it to be medical treatment, would be unlikely to decline to participate.

Most of the surviving subjects who were interviewed remembered many details of the procedures they experienced, but little or nothing that was recognizable as disclosure and solicitation of consent. They remembered being told to "take tests" (an ambiguous term in itself).[13] Two subjects did remember, along with the details of the procedures, being offered the opportunity to withdraw. Because only one military participant was located and interviewed, the Committee was not able to determine understanding on the part of the military subjects. However, given the lack of language barrier, it seems likely that the military subjects' consent to the research was voluntary and with understanding, *except for* the lack of disclosure of the radioactive nature of the tracer.

3. *"The duty and responsibility for ascertaining the quality of the consent rests upon each individual who initiates, directs or engages in the experiment. It is a personal duty and responsibility which may not be delegated to another with impunity."*

Because of the language barrier and the social and cultural organization of Alaska Native villages, it was necessary and appropriate for the researchers to make use of village authority figures as interpreters and to assist in the recruitment of subjects. However, some aspects of the process of subject recruitment and information-giving were troubling to the Committee. The difficulties of adequate translation in this setting have already been described; in addition, however, the village elders were essentially given unsupervised authority for subject recruitment, under circumstances where no researchers were available to monitor the process, to supplement and correct the information given, or to answer questions before villagers lmade the commitment to participate. Thus, through the recruitment process the researchers delegated to others to others their nondelegable duty to obtain the subjects' informed consent, and compromised the subjects's understanding and voluntariness. It also appears that the interpreters conveyed initial information to prospective subjects without having the researchers present to make corrections or answer questions. Since the researchers did communicate with enrolled subjects through

---

[12]Based on testimony at the hearings and in villages, subjects viewed themselves as obligated participants from the moment they volunteered or were identified by the elder or interpreter, before they had any opportunity to hear details of the research from the researchers.

[13]The Committee heard historical testimony (from Robert Fortuine, physician and historian) that health services were provided to Alaska Natives by the U.S. Department of Education before the Bureau of Indian Affairs took on that function. Thus, it seems logical to speculate that Alaskan Natives who recalled "taking tests" and "the studies" were not merely using simple terms for research or experiment, but were also reflecting their confusion at the time over the blurring together of education, health care, and non-therapeutic research. The researchers' emphasis on teaching and learning, while intended to convey respect for Native survival skills, probably added to the confusion. Dr. Rodahl himself stated very emphatically, "We were scientists; we were researchers." His clarity of purpose did not seem to be transmitted to the Alaska Native subjects.

interpreters during the conduct of the research to explain what the subjects were expected to do, their failure to take part in the recruitment process is noteworthy. Moreover, although the physicians needed interpreters to communicate and the use of village authority figures accorded respect to village leadership and customs, the approach could convey the impression that participation was expected for the good of the village. Thus, while the Committee believes the elders had to be involved, that involvement should have been carefully orchestrated, with the full involvement of the researchers, to overcome language problems (both of translation and of conveying concepts like research verson treatment, and radiation) and to help balance the authority of the elders with information and assurances that nonparticipation was acceptable. What happened instead is that the researchers apparently relied on the elders to bring back volunteers, who were assumed to be willing and were then told what to do, carefully and in some detial. The same amount of detail and oversight should have accompanied the recruitment process.

The notion of seeking informed consent from individuals was certainly not foreign to Alaska Natives at the time. Although Native peoples, including Alaska Natives, emphasize cooperation and community, individuals still have important rights and interests (see, for instance, Ijsselmuiden and Faden, 1992). Moreover, Alaska Natives were not outside American legal structure: even though Alaska did not become a state until 1959, Alaska Natives in the Territory of Alaska were United States citizens at the time of the research, according to the Compiled Laws of Alaska §20-1-7 (1952). Thus, they enjoyed all the rights and interests of American citizens. The Committee agrees that it was appropriate for the researchers to ask village elders to begin the process of subject recruitment, but after this politicall courtesy the researchers themselves should have obtained the voluntary and informed consent of individual subjects, using village elders to assist in translation of necessary information during the consent process.

The Common Rule and the AAL Research: Looking Back from the 1990s

Although the basic requirements of the Nuremberg Code as applied to the AAL research are still applicable today, under the terms of the Common Rule current concerns give even more emphasis to protecting vulnerable minorities or regions wielding relatively little political influence. Much of this increased attention to and understanding of justice issues is relatively new; many of the questions now framed in justice terms have not yet been well answered (see Beauchamp and Childress, 1994; National Commission for the Protection of Human Subjects of Biomedical and Behaviourial Research, 1979).

Justice in research design, according to the *Belmont Report* and the regulations that were the precursors to the federal Common Rule, is addressed primarily by the equitable selection of research subjects. Research subjects must be treated individually as capable adult decision-makers, deserving of respect, equal treatment, and fair dealing. Where populations of subjects are vulnerable to exploitation because of cultural differences or economic or social inequalities, special care must be taken to guard against their exploitation as subjects of research. Sometimes minorities are continually sought as research subjects for reasons of administrative convenience, such as when they are readily available in settings where research is ongoing. Social justice requires that researchers consider the appropriateness of adding to the cumulative research burden on any particular person and groups. The federal Common Rule, which applies to Department of Defense research, similarly requires that the selection of subjects be equitable, taking into account the purposes of the research, the setting in which it is conducted, and the special problems of involving vulnerable populations such as economically or educationally disadvantaged persons (Common Rule §111(a)(3)).

The perspective of justice becomes clearest in examining the context in which the AAL research took place as a whole, rather than the thyroid experiment in isolation. Although the principle of justice had been invoked in connection with human subjects research prior to the 1950s, at the time of the AAL research the perception of injustice resulting from social biases against minority populations was not well developed. In retrospect, a variety of activities over many decades seem to have made use of the lands and peoples of Alaska as means to the ends of others, with insufficient consideration of Alaska Natives as autonomous persons with interests of their own. Alaska Natives today believe that they and their lands have been unduly burdened, not only by research, but by a variety of other historical impositions.

The cultural characteristics of the Alaska Natives and the researchers, combined with the social and historical context at the time of the AAL research, made it highly likely that Alaska Natives would submit to the authority of researchers, and that the researchers would take acquiescence as consent. The general context of multiple interventions experienced by Alaska Natives for a variety of reasons by a variety of authorities ensured confusion between research and treatment, and bolstered the authority of researchers to dispense benefits as they chose. Indeed, during testimony to the Committee some Alaska Natives who had been subjects of medical research had difficulty recalling which of the studies they had participated in was under discussion.

At the time of the AAL research, villages both welcomed strangers and had grown accustomed to impositions by outside figures in authority. Alaska Natives considered disagreement impolite, and religious, federal, and military visitors expected acquiescence from them. The Committee found that, with the exception of Fort Yukon (which had a private hospital), visits by physicians of any kind to the rural Native villages of this study were rare in the mid-1950s, and that physicians were welcomed with open arms.

Language was clearly a barrier to comprehension in the AAL study. One of the study participants from Arctic Village told the Committee that she and the members of her family, including the tribal leader, spoke no English at the time, so they just nodded at everything the doctor said. Some village council members and tribal elders had begun to speak English as a second language at the time, but effective communication was still difficult. VanStone (1956) wrote on the subject of contacts between the U.S. Air Force and Alaska Natives:

> Problems of understanding naturally arise between peoples who have been brought up in different language systems, even though one party may have a working knowledge of the other's language. Military personnel will find speakers of English in every Eskimo village in Alaska. In most of the villages of northwest Alaska, a large percentage of the population speaks and understands some English. Only the older people have no working knowledge of the language. It should be kept in mind, however, that even when a seemingly successful conversation is being carried on in English, there are many opportunities for mis-understandings to arise. When a person is

> having difficulty in making himself understood in a language that is not his own, the natural tendency is to agree to whatever assistance the other party to the conversation is giving. It is a very easy matter to put words into another person's mouth as a part of the painful process of trying to help him express himself. Thus, military men attempting to get directions and other types of information from Eskimos should speak slowly and distinctly. The urge to push a conversation along desired lines, though sometimes irresistible, is very apt to lead to misunderstandings, since the Eskimo will not have time to order his thoughts in the foreign language and express himself clearly.

The problem of language barriers was also reported more recently by Morrow (1992), who found that in courtroom and legal situations Alaska Native Yup'ik defendants almost always answered the court's scripted questions with predictable "correct" responses, even when it was made apparent that the person did not understand several of the points to which he had agreed. According to Chance (1961), the traditionally passive role of Native women is reflected in responses to medical questioning where they defer to men. He also noted that owing to a cultural emphasis on self-reliance, anxiety is not permitted to be expressed overtly. In the Committee's view, Dr. Rodahl and other AAL researchers genuinely seem to have believed that the villagers' welcoming attitude indicated eager cooperation, and that their relationship with the subjects was one of trust and respect. Although this belief may have been paternalistic, it was commonly held (not only in the Territory of Alaska, but among many cross-cultural researchers at the time), and more culturally astute anthropological guidance was not available to the U.S. Air Force until at least 1956 (VanStone, 1956), as the AAL research was coming to a close.

The complex contextual problems raised by cross-cultural research and research with subjects suffering significant relative economic deprivation continue to this day to give rise to violations of autonomy and justice. Some in the scientific community continue to believe that research without significant risks is justifiable, despite barriers of language and culture. However, both the Nuremberg Code and current standards dictate that if the disclosure process cannot enable subjects to make decisions based on adequate understanding, then the research should be forgone.

Another issue related to justice is whether the research itself is designed to benefit the subject population. In weighing the risks and benefits of this study, it is necessary to look at the disproportionately high numbers of Alaska Natives selected to participate. The AAL research was performed to help prepare Air Force personnel to fight and survive in a possible future war fought in the Arctic. The purpose of the research was to gather information about physiological mechanisms of cold adaptation; the expected benefits were to be directly relevant to the protection of military personnel and indirectly relevant to the protection of all American citizens. Although the disproportionate burden upon Alaska Native subjects might be justified in terms of the greater good of national security, this justification would be sufficient only if it were necessary to use Alaska Native subjects almost exclusively.

Indeed, because the research involved the adaptation to exposure to cold, the researchers argued that the Alaska Native population was uniquely suited for study. However, with hindsight there is a flaw in that logic. The purpose of the research was to determine (1) whether thyroid function changed as a result of exposure to cold and (2) whether differences in thyroid function in extreme cold resulted from long-term acclimatization to cold or from inherent racial and ethnic differences. If long-term acclimatization to cold was to be effectively distinguished from racial differences in thyroid function, the I^{131} study ought also to have included long-term residents of Alaska who were racially and ethnically different from Alaska Natives (the military subjects) were relative newcomers to the environment). When the Committee asked why white Alaskans who were long-term residents had not been included, Dr. Rodahl said that recruitment of such subjects, although desirable, was not possible. This reinforces the impression that the Alaska Natives represented a subject population of convenience. The Committee recognizes that Alaska Natives were needed as subjects to answer the research question; however, their disproportionate numbers by comparison with military personnel, and the researchers' failure to seek out indigenous non-Native subjects, emphasizes that their relative lack of understanding may have made Alaska Natives easier to recruit.

The Committee recognizes the difficulties of research design and subject selection under the circumstances in which the AAL researchers worked. The researchers seem to have considered their repeated use of a population of convenience as an exchange of benefits between two groups. However, the Alaska Natives do not now—and state that they did not then—share this understanding.

According to the *Belmont Report* and the Common Rule, today the placement of this undue burden of research participation on the Alaska Native population would require special protection for these subjects. For example, if this research were to be approved today, the researchers might be required to take special care to minimize the risks of the research and to ensure the integrity of the process of informed consent. Specifically, it would become even more important for researchers to communicate, and for subjects to understand, that they were involved in research, not treatment, and that the purpose of the research according to the research protocol was primarily to benefit military personnel. The Committee does not mean to imply that Alaska Natives should never be involved in research, or to assume that they are incapable of informed decision-making. It would certainly be wrong to deny any population the opportunity to contribute to the development of knowledge. However, when "vulnerable populations" with language and cultural barriers, like Alaska Natives, are subjects of research today, researchers face an extra burden of responsibility to balance the potential benefit from the research with the interests of, and risks to, the vulnerable population.

Finally, the Committee also considered whether the AAL researchers identified and took action to help ameliorate the endemic goiter problem in Arctic Village and Anaktuvuk Pass. The Committee could not find evidence showing that the AAL provided medical follow-up after the research, or that any explanation of the potassium iodide control experiment results was provided to the research subjects. Dr. Rodahl told the Committee that he and his collaborating physicians communicated regularly with physicians of the Alaska Native Service (ANS). He maintained that a benefit of his work was that it called attention to the problem, so that the salt supply in the local stores was replaced with iodized salt. Because of the lack of availability of historical records from the AAL and the Indian Health Service from that time period, the Committee was not able to verify the actual connection between the AAL research findings and actions by the ANS to take corrective steps in locations with endemic goiter. However, AAL information about the reduced salt intake and use of uniodized salt in those interior villages was conveyed at scientific meetings in Alaska during 1956 and in subsequent papers. No Native interviewed could remember follow-up return visits by U.S. Air Force or ANS doctors in the following years, although this does not prove that the salt supplies were not changed. The Committee was told that a few subjects of the tests in Anaktuvuk Pass subsequently had thyroidectomies; this indicated there was an awareness of the problem in the medical community and possible communication between the AAL physicians and the ANS.

If research by physicians uncovers a treatable condition, the duty of beneficence suggests that the physician researchers have some follow-up responsibility to those subjects. The provision of iodized salt, to be made available in the areas of endemic goiter, or even the attempt to educate the affected communities, would have provided a benefit that did not interfere with the research. Failure to take positive action that prevents harm is a lesser wrong than taking action that causes harm. Nonetheless, having intervened in the lives of their subjects in the course of their research, and having identified and studied the goiter problem, the AAL physician-researchers had some obligation to ameliorate the situation of their subjects.

CONCLUSION

During the period of the mid-1950s when the AAL thyroid experiments took place, the 10 points of the Nuremberg Code set standards for the conduct of human subjects research. Although the Code was established in response to an egregious abuse of scientific research and prohibits research that is unnecessarily or excessively risky or harmful, as well as research that is not designed to produce socially useful results, this does not mean that "good" research was not subject to the Code's standards. In recognition of the importance of the Code, the AMA had adopted its principles as a standard in 1947, and issues of informed consent and the Code were discussed at length by the American Association for the Advancement of Science in 1953 in the journal *Science*. By adopting the Code's main points in the Department of Defense, classified memorandum of 1953, the Air Force Surgeon General became responsible for ensuring that these consent requirements were met by the physicians under his command, including the AAL. In turn, the AAL physicians and the Air Force were responsible for ensuring that they received the voluntary and informed consent of individual subjects.

The Committee determined that because information on the nature of the I^{131} tracer was not disclosed, the military and Native subjects were not completely informed about the nature and risks of the experiments. This omission was a wrong to the subjects, even though the risk of harm was believed at the time to be nonexistent and has since been determined to be quite low. In most of the Native villages, the AAL physicians accepted as volunteers those Natives who were brought to them by village elders or other Native intermediaries without inquiring as to what the subjects had been told. The elders or other intermediaries used by the AAL physicians did not have sufficient training in medicine or science to explain the research adequately, and many of the Native witnesses for the Committee thought they were receiving a beneficial medical treatment. Minor children were used as research subjects without adequate parental consent. Few of the Alaska Native subjects understood that they were participating in research. None of the subjects, neither Alaska Natives nor military personnel, were informed that they were taking a radioactive tracer. Thus, the Committee believes that the experiments were conducted without informed consent, even according to the standards of the time.

Under current standards (represented by the Common Rule adopted by the Department of Defense), the complete experiment, its purpose, and potential risks and benefits would have to be explained to the subjects, and permission to proceed would be required (usually in writing). Thus, the AAL experimental design and consent procedure would fall short by modern informed consent standards.

The experiments raise justice concerns according to current ethical standards by virtue of the way a minority population was used. The Committee's examination strongly suggests that the Native population was used primarily because it was a convenient research pool, with an assumed understanding that the research would be conducted in exchange for medical care-giving but without full disclosure about the nature of the research. Although the Alaska Natives brought special physiological characteristics because of their long-term adaptation to the climate, long-term white Alaska residents were not sought.

The AAL researchers held a genuine belief, justifiable at the time, that the research they were conducting was both harmless and necessary. However, the Committee's examination shows that the process by which the researchers obtained consent was flawed because none of the subjects were provided with a full understanding of the nature of the research. The Alaska Native subjects also were wronged by the nature of the subject selection process and by the exploitation (even if unintentional) of language and cultural differences, which substantially reduced the likelihood that they would even understand that they were the subjects of research rather than the beneficiaries of medical treatment.

Given that the researchers and their Air Force superiors made errors of omission that were typical in the research of the times, it is important not to assign blame for these deficiencies. Instead, the appropriate response is to acknowledge their errors in the hope of ensuring that similar ones do not occur again.

4

Conclusions and Recommendations

After examining the records, analyzing the health risks, and talking with research participants as well as researchers, the Committee concludes that in all probability the AAL thyroid function study caused no physical harm to the subjects. As calculated in Chapter 2, the weighted average risk among the populations that participated in the AAL thyroid function study is about 1 in 3,000, a risk six times lower than the background thyroid cancer risk in the United States. The greatest risks (albeit small) of thyroid cancer appear in the people who received multiple I^{131} doses. In particular, the Anaktuvuk Pass females and the Arctic Village females who received multiple doses have calculated risks of 1 in 800 and 1 in 700, respectively. Because thyroid cancer is rare (about 5 cases per 100,000 population annually), the additional radiological risk is extremely low, and radiation-induced thyroid cancers caused by the AAL study would not be expected in either the Alaska Native or white military personnel who participated as research subjects. Some health benefits may have been coincidentally provided because the researchers identified endemic goiter problems in Arctic Village and Anaktuvuk Pass.

From an ethical perspective, the Committee concludes that the Alaska Natives who participated and, to a lesser extent, the military research subjects were wronged. Although the AAL thyroid function study was conducted according to generally accepted scientific and medical procedures of the times, there was a specific violation of existing ethical standards because information on the I^{131} tracer was not disclosed. Thus the Alaska Native and military subjects were not fully informed about the nature and risks of the research. This omission was wrong, even though the risk of harm was believed at the time to be nonexistent and has since been determined to be extremely low.

The blurring of medical care and research activities, as well as cultural differences, compromised the ability of Alaska Native subjects to recognize not only that they were participating in research, but that refusal to participate would not harm their health. The language barrier between researchers and study subjects in the Native villages and the use of Native interpreters who were not scientifically trained prevented participants from gaining a complete understanding of the research and its risks. In addition, some study subjects should have been excluded on the basis on the basis of age or other physical conditions.

The study's flaws are not attributable solely to the AAL researchers in the field, but are shared by their superiors in the U.S. Air Force and the Department of Defense chain of command who approved the research or were aware of it. Indeed, what we today identify as inappropriate was common for the times. It should be noted that during the 1950s and later many researchers viewed obtaining consent as essential only for obviously risk-bearing research, because of the circumstances that produced the Nuremberg Code. However, neither the Code by its terms nor the American Medical Association (AMA) or DOD requirements based on it contain such a caveat.

It is important to emphasize that while the Committee believes it is inappropriate to place individual blame for the flaws of the AAL study, it is essential for the government to acknowledge that wrongs were done. The researchers were conscientious scientists who held genuine belief, justified at the time, that their research was both harmless and important. The lack of emphasis on autonomy and informed consent, and the lack of cultural sensitivity, were standard errors of the time. It is the Committee's hope that acknowledgment of these wrongs will reduce the likelihood of similar wrongs in the future.

The Committee heard repeated concerns about the history of abuse of the Native people by the incoming white culture. In the Alaska Natives' view, the AAL study was a small but characteristic event in the ongoing tradition of using the Native peoples for the benefit of others with little regard for their interests. Although the Committee was limited by charge and resources from considering this wider context in any depth, its recommendations are not unrelated to this underlying concern. As explained in a review of the social and psychological stress faced by veterans of above-ground nuclear testing and similar nuclear weapons-connected events conducted by the government, there is a relationship between trust and acknowledgment of error (Garcia, 1994, p. 654): "[such veterans] . . . must deal with the possibility that their lives were undermined without apology, acknowledgment of error, or recognition of their service to their country."

The Department of Energy, in a parallel activity investigating the use of human subjects in radiation research, acknowledges the importance of dealing with past wrongs directly and openly. As stated in a 1995 report (Department of Energy, 1995):

> Over the past several decades, the American people's trust in our institutions of government has greatly eroded. Many complex factors have contributed to this erosion, not least among them the secrecy associated with our Cold War nuclear competition with the Soviet Union. Without judging the historical necessity of secrecy, and in recognition that even today some activities require national security classification, it is a fact that the ability of the Government to perform its post-Cold War missions is greatly impeded by pervasive public distrust of its motives and competence. The commitment to openness, of which this project is a very visible element, is a deliberate effort to rebuild that basic level of trust between the American people and their government that is necessary for a democracy to function.

The Alaska Natives who spoke to the Committee were clearly frustrated by the lack of communication and lack of acknowledgment. Until this Committee held its public meeting, beliefs about how the government conducted itself had not been given a forum for expression. As a result, Alaska Natives are dealing with unresolved burdens in regard to trust and justice. However belatedly, the Air Force, U.S. health organizations, or the Congress could redress the wrong of failure to obtain informed consent during the AAL thyroid function study with information now. The ongoing provision of meaningful information could provide surviving subjects, their families, and their villages with better understanding of the true magnitude of the risks and possible consequences of the research. This demonstration of respect for their autonomy and concern for their well-being could help restore a sense of control to the Native populations over their own health decisions and hopefully increase the level of trust. The present inquiry affords the federal government a singular opportunity to give audience to urgent and long-held Native concerns. In this spirit, the Committee recommends the following:

> *(1) The government and the Air Force should acknowledge responsibility for wrongs done in the course of the AAL thyroid function study in the hopes of ensuring that similar problems do not occur in the future, and they should address these wrongs by undertaking the following actions:*
>
> > *(a) The Air Force should endeavor to contact all living subjects or their immediate families and provide records to them of their AAL research participation in the I^{131} experiments. The Air Force should also continue to search for records of the AAL that would identify the six U.S. Army subjects and six Point Hope subjects who were not named in the Air Force report of the study, and to locate the Air Force and Army subjects named in the study.*
> >
> > *(b) In the process of contacting subjects and subjects' families, the Air Force should disseminate the Committee's report and other available information on human medical experimentation conducted by the AAL in the period 1948-1967 to appropriate health care providers, tribal governments, and other key figures in the relevant Alaska Native villages.*

This dissemination of information could be accomplished in the six affected Native villages by having a group of medical and ethical experts provide a briefing on the Committee's report at a town meeting and answer questions related to the AAL research. A concise, readable summary of such information (preferably bilingual) also should be prepared. Even though material related to the AAL studies is available in larger Alaska public libraries, the lack of knowledge about studies, tests, and research conducted by that government facility has haunted participants and their families, some of whom took part in even the most benign of studies.

> *(2) U.S. government and Alaska state health organizations, under U.S. government auspices, could complement the efforts of the Air Force by conducting related public health education programs facilitated by Native experts focused on conveying information about patients' rights in any therapeutic or research situation and medical information about exposure to radiation. Such a process will enable Native experts, clinics, and physicians to provide accurate information to their communities and ease fears.*

(3) If Congress considers legislation[1] *to redress any wrongs or harms done to human subjects of government radiation research where informed consent was not obtained, the Committee believes Congress should consider including the subjects of the AAL thyroid function study.*

The last question posed to the Committee was whether or not follow-up surveys should have been performed to ensure that the participants suffered no long-term ill effects from the experiments and whether medical care was needed. Extending that to the present, the Committee also considered whether medical follow-up is warranted today based on our current understanding of the risks involved. Such medical follow-up would focus on the major negative health effects associated with I^{131} exposure—disorders of the thyroid gland, including thyroid cancer and the development of thyroid nodules. (There is no evidence of a link between I^{131} exposure and skin disorders, a concern raised by some study participants at the public hearing.)

Because the dosages of radioactive iodine used in the AAL study were thought to be harmless at the time and there were no guidelines requiring follow-up for diagnostic doses, no follow-up would have been indicated. In reexamining the doses of radioisotope ingested by the research subjects, the Committee concludes that there is no justification based on risk for medical follow-up. The risk, even among research subjects who received multiple doses, is substantially lower than the background incidence of thyroid cancer in the United States.

(4) Although medical follow-up based on the calculated risk values is not warranted, the U.S. Air Force should provide medical follow-up to those participants who were under age 20 at the time of the AAL study since those participants will be at risk for the longest period of time. Such follow-up would provide assurance that these participants suffered no long-term physical ill effects.

---

[1]Various laws have some bearing on issues in this study. For instance, common law provides that governments cannot be sued by their subjects (*Feres v. United States*, 340 U.S.C. 135, 139 (1950)). The Federal Tort Claims Act (28 U.S.C. v§2674 et seq. (1988)) authorizes federal liability for compensatory damages in limited circumstances, but not when the injury results from the exercise of discretionary judgment by government officials—an exception that has barred recovery for uranium miners' on-the-job exposure (*Begay v. United States*, 768 F.2d 1059 (9th Cir. 1985)) and for radiation exposure at nuclear test sites (e.g. *Prescott v. United States* (D.Nev. July 21, 1994)), and that seems clearly to apply in the case of medical and scientific research. Legislation exists to compensate some persons exposed to radiation by the government—the Veterans Radiation Exposure Compensation Act (42 U.S.C. §2210nt), and the Compact of Free Association with the Marshall Islands (48 U.S.C. §1681nt).

However, none of these existing statutes addresses the subjects of the AAL study or, for that matter, of other Cold War-era radiation experiments under investigation by other bodies, notably the President's Advisory Committee on Human Radiation Experiments. If such legislation is considered, the language must be carefully chosen to ensure that it encompasses all of the human subjects who, in the eyes of Congress, deserve compensation for wrong and/or harms. Two related bills were considered in the 103rd Congress, but neither passed out of committee. As this report goes to publication, the Radiation Experimentation Compensation Act (H.R. 2463) had been reintroduced in the 104th Congress and referred to the Judiciary Committee; the Radiation Experimentation Victims Act was being redrafted, with plans to reintroduce it.

Such follow-up should be at government expense. It might include a review of the subjects' medical history and a physical medical examination with special attention to physical complaints, ailments, or physiological changes or disease that could be related to the AAL thyroid function study, specifically assessment of the condition of the participants' thyroids for the occurrence of thyroid cancer or nodules. If a thyroid disease, including cancer, were to be found, there would be no way of determining whether participation in the AAL study had any role in causing it. Nonetheless, if thyroid disease were found the federal government should be responsible for all expenses associated with treatment. In the remote possibility that thyroid disease is found for those under 20, the Air Force should then reassess the issue and decide if more subjects should receive medical follow-up.

The Committee recognizes that its basic conclusion—that the subjects of the AAL thyroid function study were wronged but not harmed—may prove controversial. Some will claim that the Committee's calculations are incorrect and that the risk is higher. Others will believe that the Committee failed to go far enough in suggesting ways to right the wrongs. Some will say that the Committee failed to understand the climate of the times—the intensity of Cold War pressures and national security concerns and the fact that many researchers truly did not believe that the Nuremberg Code applied to benign human subjects research. They may claim that the Committee was swayed by the clarity that only hindsight brings.

The Committee believes that these various perspectives arise from concern for the people involved, both the researchers and their superiors and the research subjects. It recognizes that some subjectivity is inherent in this type of analysis and that honest differences of opinion can occur. Still, the Committee is convinced that its position is defensible, sensible, and ethical. The risk analysis in this report is based on the best epidemiology and dosimetry available. It is, if anything, conservative; risks may actually be smaller than expressed. The Committee's position acknowledges the flaws of the AAL thyroid function study within the context of history, while not placing blame on those who conducted the research using what they perceived to be harmless methods in pursuit of justifiable goals.

References

Advisory Committee on Human Radiation Experiments. 1995. Final Report. U.S. Government Printing Office, Pittsburgh, PA.

Ad hoc Committee on Polar Biomedical Research, Polar Research Board. 1982. Polar Biomedical Research: Appendix—Polar Medicine—A Literature Review. National Academy Press, Washington, DC.

Ahmed, S.R., and S.M. Shalet. 1985. Gonadal damage due to radioactive iodine (I^{131}) treatment for thyroid carcinoma. Postgrad Med J 61:361-2.

American Cancer Society, Inc. 1995. Cancer Facts & Figures-1993. American Cancer Society, Inc., Atlanta, GA.

Arctic Aeromedical Laboratory. 1957. Report of Operations, July 1, 1956 to December 31, 1955. U.S. Air Force, Alaska Air Command, Fairbanks, AK.

Baeumler, G.R. 1986. Radioactive iodine uptake by breasts (letter). J Nucl Med 27:149-51.

Bagchi, N., T.R. Brown, and P.F. Parish. 1990. Thyroid dysfunction in adults over age 55 years. A study in an urban US Community. Arch Intern Med 150: 785-7.

Bauer, F., W.E. Goodwin, R.L. Libby, and B. Cassen. 1953. The diagnosis of morphologic abnormalities of the human thyroid gland by means of I^{131}. Radiology 61:935-37.

Baker, T.G. 1971. Radiosensitivity of mammalian oocytes with particular reference to the human female. Am J Obstet Gynecol 110:746-61.

Beauchamp, T., and J. Childress. 1994. Principles of Biomedical Ethics, 4th ed., chapter 6. Oxford University Press, New York.

Blahd, W.H., 1965. In Nuclear Medicine, McGraw-Hill Book Company, New York.

Brown, H., and Hatcher, C.H. 1953. Acclimatization to Cold, Metabolic and Vascular Studies. Proceedings XIX International Physiology Congress, p. 240.

Burch, E. 1976. *"Nunamiut Concept and the Standardization of Error in Contributions to Anthropology: The Interior Peoples of Northern Alaska*. Edwin S. Hall, Jr., editor. Mercury Series. Archaeological Survey Paper 49. Canada National Museum of Man, Ottawa, pp. 52-97.

Chance, N. 1961. Conceptual and Methodological Problems in Cross-Cultural Health Research. Proceedings of Twelfth Alaskan Science Conference. Alaska Division of American Association for the Advancement of Science, pp. 14-6.

Chapman, E.M. 1983. History of the discovery and early use of radioactive iodine. JAMA, v. 250: p. 2042-4.

Chopra, I.J. 1991. Nature, sources, and relative biologic significance of circulating thyroid hormones. In Werner and Ingbar's "The Thyroid" (L.E. Braverman and R.D. Utiger, eds), J.B. Lippincott Company, Philadelphia, p. 127143.

Chopra, I.J., and D.H. Solomon. 1979. Thyroid function tests and their alterations by drugs. In The Thyroid, Physiology and Treatment of Disease (J.M. Hershman and G.A. Bray, eds.), Pergamon Press, New York, p. 105-137.

Clark, D.E. 1956. Proceedings of the Conference on Radioiodine, November 5 and 6, 1956, Chicago, Illinois. University of Chicago.

Cooper, D.S., J.L. Cevallos, R. Houston, N. Chagnon, and P.W. Ladenson. 1993. The thyroid status of the Yanomano Indians of southern Venezuela. J Clin Endocrinol Metab 77: 878-80.

Cox, P.H., J.G.M., Klinj, M. Bontebal, and D.H.W. Schonfeld. 1990. Uterine radiation dose from open sources: The potential for underestimation. Eur J Nucl Med 17:94-5.

Department of Energy. 1995. *Human Radiation Experiments: The Department of Energy Roadmap to the Story and the Records*. Natl. Tech. Info. Service, Springfield, Virginia.

Department of Health, Education, and Welfare, Food and Drug Administration, Radioactive Drugs and Radioactive Biological Products. July 25, 1975. Fed Reg 40:31308-9.

Department of Health, Education, and Welfare. National Commission for the Protection of Human Subjects of Biomedical and Behavioral Research. 1979. Belmont Report.

Department of Health and Human Services, Food and Drug Administration. 1990. Radioactive Drugs for Certain Research Uses. 21 CFR 361.1, revised April 1993.

Dick, M., and F. Watson. 1980. Prevalent low serum thyroxine-binding globulin level in Western Australian aborigines: its effect on thyroid function tests. Med J Aust 1: 1156-8.

Drury, H., M. Hall, R. Glisczinski R, and J. Spence. 1956. Nutritional Survey at Anaktuvuk Pass. Proceedings of the VII Alaskan Science Conference, pp. 95-6.

Dydek, G.J., and P.W. Blue. 1988. Human breast milk excretion of I^{131} following diagnostic and therapeutic administration to a lactating patient with Graves' disease. J Nucl Med 29:407-10.

Edmonds, C.J., and T. Smith. 1986. The long-term hazards of the treatment of thyroid cancer with radioiodine. Br J Radiol 59:45-51.

Faden, R., T. Beauchamp, and N. King. 1986. *A History and Theory of Informed Consent*, Oxford University Press, New York.

Federal Common Rule §111(a)(1)-(2). 1991. Federal Policy for the Protection of Human Subjects, Fed Reg, vol. 56, 116. pp: 28016-17.

Garcia, B. 1994. Social-psychological dilemmas and coping of atomic veterans. Am J Orthopsychiatry-64(4):651-55.

Green, H.G. et al. 1971. Cretinism associated with maternal sodium $iodide^{131}$ therapy during pregnancy. Am J Dis Child 122:247-9.

Greenspan, F.S., B. Rapoport. 1983. Thyroid gland, In Basic and Clinical Endocrinology (F.S. Greenspan and P.H. Forsham eds.). Lange Medical Publications, Los Altos, California, p. 120-186.

Hanson, W.C. 1967. $Cesium^{137}$ in Alaska lichens, Caribou and Eskimos. Health Phys 13:383-9.

Hanson, W.C. 1971. ^{137}Cs: seasonal patterns in Native residents of three contrasting Alaskan villages. Health Phys v22:39-42.

Hanson, W.C. 1982. ^{137}Cs concentrations in Northern Alaska Eskimos, 1962-79: Effects of Ecological, Cultural and Political Factors. Health Phy v42:433-47.

Hanson, W.C. 1994. Radioactive Contamination in Arctic Tundra Ecosystems. In: Workshop on Arctic Contamination, May 2-7, 1993, Anchorage, Alaska. Interagency Arctic Research Policy Committee, pp.198-206.

Hedrick, W.R., R.N. DiSimone, and R.L. Keen. 1987. Radiation dosimetry from breast milk excretion of I^{123}. J Nucl Med 28:544-45.

Hendee, W.R., 1993. History, current status, and trends of radiation protection standards. Med Phy 20:1303-14.

Holm, L.E., K.E. Wiklund, G.E. Lundell, N.A. Bergman, G. Bjelkengren, E.S. Cederquist, U.B.C. Ericsson, L.G. Larsson, M.E. Lidberg, R.S. Lindberg, H.V. Wicklund, and J.D. Boice, Jr. 1988. Thyroid cancer after diagnostic doses of $iodine^{131}$: A retrospective cohort study. JNCI 80:1132-38.

Hopkins, C., C. Dotter, and H. Griswold. 1958. Longitudinal Medical-Anthropological Observations of the Northwestern Coastal Eskimo of Alaska, with Special Emphasis on Cardiovascular Disease. Proceedings of Ninth Alaskan Science Conference. Alaska Division of American Association for the Advancement of Science, pp. 120-131.

Ijsselmuiden, C.B., and R.R. Faden. 1992. Research and Informed Consent in Africa—Another Look. New England J Med v. 326 pp. 830-3.

Journal of the American Medical Association (JAMA). 1958. JAMA published the AMA's 1957 Principles of Medical Ethics and Opinions and Reports of the Judicial Council on December 28, 1946.

Lanier, A.P., J.J. Kelly, B. Smith, C. Amadon, R. Harpster, H. Peters, and H. Tanttila. 1994. Cancer in the Alaska Native population: Eskimo, Aleut, and Indian, incidence and trends 1969-1988. Alaska Med, v36(1):5-87.

Laughlin, W. 1957. Blood Groups of the Anaqtuavik Eskimos, Alaska. Arctic Aeromedical Laboratory, Ladd Air Force Base, Alaska, Technical Report 57-5.

Lawes, S.C. 1992. I^{123} excretion in breast milk—additional data. Nucl Med Comm 13:570-3.

Leblond, G., and J. Gross 1943. Thyroidectomy on resistance to low environmental temperature. Endocrinology. v33:155.

Leblond, G., J. Gross, W. Peacock, and R. Evans. 1944. Metabolism of radio-iodine in the thyroids of rats exposed to high or low temperature. Am J Phy v140:671.

Marsh, G. 1954. Comparative Survey of the Eskimo-Aleut Religions. Anthropological Papers of the University of Alaska. Volume 5(1).

Meehan, J. 1955. Individual and Racial Variations in a Vascular Response to Cold Stimulus. Arctic Aeromedical Laboratory, Ladd Air Force Base, Alaska, Project 7-7953, Report No. 1.

Milan, F. 1962. The Acculturation of the Contemporary Eskimo of Wainwright, Alaska. Anthropological Papers of the University of Alaska. Volume 11(2).

Morrow, P. 1992. Culture and communication in the Alaskan courtroom—A place to be made to talk. Arctic J US 6:65-70.

Mossman, K.L. 1992. Radiation protection of human subjects in research protocols. In: Mossman KL and Mill WA eds. *The Biological Basis of Radiation Protection Practice*. Williams and Wilkins, Baltimore, pp. 255-61.

NAS (National Academy of Sciences). 1990. BEIR V. *Health Effects of Exposure to Low Levels of Ionizing Radiation*. National Academy Press, Washington, DC.

NCRP (National Council on Radiation Protection and Measurements). 1989. *Exposure of the U.S.Population from Diagnostic Medical Radiation*. NCRP Report No. 100. NCRP, Bethesda, MD.

NCRP (National Council on Radiation Protection and Measurements). 1985. *Induction of Thyroid Cancer by Ionizing Radiation*. NCRP Report No. 80. NCRP, Bethesda, MD.

National Commission for the Protection of Human Subjects of Biomedical and Behavioral Research. 1979. The Belmont Report: Ethical Principles and Guidelines for the Protection of Human Subjects of Research. Washington, D.C.: Public Health Service.

Nuremberg Code. 1946. Trials of War Criminals before the Nuremberg Military Tribunals under Control Council Law No. 10:Nuremberg, October 1946-April 1949. Washington. Government Printing Office (n.d.), vol. 2, pp:181-2.

President's Advisory Committee on Human Radiation Experiments Interim Report. October 21, 1994. 41 pp. plus appendices.

Quimby, E.H., S. Feitelberg, and S. Silver. 1958. In Radioactive Isotopes in Clinical Practice, Lea and Felsiger, Philadelphia.

Rall, J.E. 1956. The role of radioactive iodine in the diagnosis of thyroid disease. Am J Med 20:719-31.

Rall, J.E. 1957. Radiation and the medical profession (editorial). AMA Arch Intern Med: 100:347-52.

Reed, H.L., D. Brice, K.M. Mohamed Shakir, K.D. Burman, M.M. D'Alesandro, and J.T. O'Brian 1990. Decreased Free Fraction of Thyroid Hormones After Prolonged Antarctic Residence. J. Appl. Physiol. 69:1467-90.

Reed, H.L., E.D. Silverman, K.M. Mohamed Shakir, R. Dons, K.D. Burman, and J.T. O'Brian. 1990. Changes in serum triiodothyronine (T_3) kinetics after prolonged Antarctic residence: The polar T_3 syndrome. J Clin Endocrinol Metab 70:965-74.

Rodahl, K. 1952. Basal Metabolism of the Eskimo. Arctic Aeromedical Laboratory, Ladd Air Force Base, Alaska Project. 22-1301-001, part 2.

Rodahl, K. 1954. Studies on the Blood and Blood Pressure in the Eskimo and the Significance of Ketosis Under Arctic Conditions. Norsk Polar Institute, Skrifter, No. 102.

Rodahl, K. 1962. *Last of the Few*. Harper and Row, New York.

Rodahl, K. 1957. Human Acclimatization to Cold. Arctic Aeromedical Laboratory Technical Report 57-21. Ladd Air Force Base.

Rodahl, K., and G. Bang. 1956a. Endemic Goiter in Alaska. Arctic Aeromedical Laboratory, Ladd Air Force Base, Alaska, Technical Report AAL-TN-56-9.

Rodahl, K., and G. Bang. 1956b. Endemic Goiter in Alaska. Proceedings of Seventh Alaskan Science Conference. Alaska Division of American Association for the Advancement of Science, pp. 101-2.

Rodahl, K., and G. Bang. 1957. Thyroid Activity in Men Exposed to Cold. Arctic Aeromedical Laboratory, Ladd Air Force Base, Alaska, Technical Report 57-36.

Rodahl, K., and D. Rennie. 1957. Comparative Sweat Rates of Eskimos and Caucasians Under Controlled Conditions. Arctic Aeromedical Laboratory, Ladd Air Force Base, Alaska, Project 8-7951-18, Report No. 7.

Romney, B., E.L. Kickoloff, and P.D. Easer. 1989. Excretion of radioiodine in breast milk (editorial). J Nucl Med 30:124-6.

Ron, E., J. Lubin, R. Shore, et al. 1995. Thyroid cancer after exposure to external radiation: A pooled analysis of seven studies. *Radiat Res* 141:259-77.

Schectman, J.M., G.A. Kallenberg, R.P. Hirsch, and R.J. Schumacher. 1991. Report of an association between race and thyroid stimulating hormone level. Am i Public Health 81:505-6,

Science Magazine. 1953. *The Problem of Experimentation on Human Beings*. February 27, 1953.

Shepard, T.H. 1967. Onset of function in the human fetal thyroid: Biochemical and radioautographic studies from organ culture. J Clin Endocrinol Metab 27:945-58.

Shore, R. 1992. Issues and epidemiological evidence regarding radiation-induced thyroid cancer. *Radiat Res* 131:98-111.

Shore, R.E., L.H. Hempelmann, and E.D. Woodard. 1986. Carcinogenic effects of radiation on the human thyroid gland. In: AC Upton, RE Albert, FJ Burns, and RE Shore, eds. *Radiation Carcinogenesis* Elsevier, New York, pp. 293-309.

Simpson, C.L., L.H. Hempelmann, and L.M. Fuller. 1955. Neoplasia in children treated with x-rays in infancy for thymic enlargement. Radiology 64:840-5.

Slater, P.E., J.D. Kark, Y. Friedlander, and N.A. Kaufman. 1984. Determinants of thyroid hormone and thyroid lipid interrelationships in Jerusalem. Ir i Med Sci 20: 1158-63.

Spencer, R. 1984. North Alaska Coast Eskimo. In: *Handbook of North American Indians, vol.5. Arctic.* Smithsonian Institution, Washington, DC, pp. 320-37.

Stabin, M., E.E. Watson, C.S. Marcus, and R. Salk. 1991. "Radiation dosimetry for the adult female and fetus from I131 administration in hyperthyroidism." J Nucl Med; 31:808-813.

Stutzman, C.D., D.M. Nelson, and L.P. Lanier. 1986. Estimates of cancer incidence in Alaskan Natives due to exposure to global radioactive fallout from atmospheric nuclear weapons testing. Alaska Med 28:53-63.

Therien, M. 1949. Contribution to the physiology of cold acclimatization. Laval Medicine, vol. 4, November 8-9.

Tkachev, A.V., E.B. Ramenskaya, and J.R. Bojko. 1991. Dynamics of hormone and metabolic state in polar inhabitants depend on daylight duration. Arctic Med Res 50 (Suppl 6):152-5.

U.S. Air Force. Undated. (U.S. Air Force Alaska Air Command, 1955a, 1955b, 1956a, 1956b, 1957a, 1957b, 1958a).

U.S. Air Force. 1955a. History of the Arctic Aeromedical Laboratory, 1 January 1955-30 June 1955. Alaskan Air Command. Elmendorf Air Force Base, Anchorage, Alaska.

U.S. Air Force. 1955b. History of the Arctic Aeromedical Laboratory, 1 July 1955-31 December 1955. Alaskan Air Command, Elmendorf Air Force Base, Anchorage, Alaska.

U.S. Air Force. 1956a. History of the Arctic Aeromedical Laboratory, 1 January 1956-30 June 1956. Alaskan Air Command, Elmendorf Air Force Base, Anchorage, Alaska.

U.S. Air Force. 1956b. History of the Arctic Aeromedical Laboratory, 1 July 1956-31 December 1956. Alaskan Air Command, Elmendorf Air Force Base, Anchorage, Alaska.

U.S. Air Force. 1957a. History of the Arctic Aeromedical Laboratory, 1 January 1957-30 June 1957. Alaskan Air Command, Elmendorf Air Force Base, Anchorage, Alaska.

U.S. Air Force. 1957b. History of the Arctic Aeromedical Laboratory, 1 July 1957-31 December 1957. Alaskan Air Command, Elmendorf Air Force Base, Anchorage, Alaska.

U.S. Air Force. 1958a. History of the Arctic Aeromedical Laboratory, 1 January 1958-30 June 1958. Alaskan Air Command, Elmendorf Air Force Base, Anchorage, Alaska.

VanStone, J. 1956. Report on Air Force Eskimo Contacts. Arctic Aeromedical Laboratory, Ladd Air Force Base, Alaska, Technical Report 56-15.

World Medical Association. 1961. Declaration of Helsinki.

Appendix A
Thyroid Function in Health and Disease

THE HUMAN THYROID

The discussion that follows describes in technical terms the role and function of the thyroid gland in health and disease, and provides a brief description of racial differences that have been observed. The purpose is to put into perspective why the thyroid was a likely study candidate in the 1950s for research in acclimatization to cold in the AAL thyroid study (Rodahl and Bang, 1957) and why radioactive iodine was utilized.

The thyroid gland is one of several glands in the body that makes and secretes hormones necessary for life. The hormones entering the bloodstream have biochemical effects on certain targeted body tissues. In humans the thyroid consists of two lobes in the lower part of the neck, on either side of the windpipe (trachea). These are the left and right lobes, which in turn are connected by a thin strand of thyroid called an isthmus that is located at or near the Adam's apple in the neck (cricoid cartilage). The level and activity of the thyroid is regulated by thyroid-stimulating hormone (TSH), which is secreted by the anterior pituitary gland, a peanut-sized organ located just beneath the brain that regulates various aspects of the body's growth, development, and functioning. The thyroid hormones, thyroxine (T_4) and 3,5,3'-triiodothyronine (T_3), are important to growth and development of the fetus and child, and they regulate metabolic processes in essentially all tissues throughout life. The thyroid gland[1] is the only known source of T_4. T_4 is also metabolized in some peripheral tissues to a more active hormone, 3,5,3'-T_3 by one enzyme or to a less active substance, 3,3',5'-T_3 (reverse T_3, rT_3) by the action of a different enzyme (Chopra, 1991).

[1]The thyroid gland also produces some T_3, but the bulk ($\sim$75 percent) of T_3 is produced outside of the thyroid in extrathyroidal tissues (e.g., liver and kidney) by enzymatic conversion of T_4 to T_3. T_3 is some three to five times more active than T_4 in two important effects of thyroid hormones: increase of oxygen consumption and suppression of TSH secretion by the anterior pituitary gland. Reverse T_3 is only 1 percent as active as T_4 in these two actions; its biological significance is not well understood at this time.

ROLE OF IODINE IN THYROID FUNCTION AND DIAGNOSTIC TESTS

Iodine is a critical element in the normal production of thyroid hormones. The iodine is derived from iodide included in the diet. A typical dietary intake of iodide in the United States is approximately 500 micrograms (μg) per day, 200-300 μg in the eastern states and 500-750 μg in California. The main sources of iodide are water, bread, salt, kelp seaweed, and certain medicines. Iodide is nearly completely absorbed from the gastrointestinal tract and enters the inorganic pool in the extracellular fluid. The thyroid gland picks up some 75 μg of iodide per day for synthesis of thyroid hormones, and the rest is cleared (excreted) by the kidneys. When the dietary intake of iodide is reduced, the thyroid increases its uptake of iodide in the extracellular fluid, and vice versa. This is done in order to maintain normal amounts of thyroid hormone synthesized by the thyroid. Understanding this important concept of the regulation of thyroidal uptake of iodide became the basis for using thyroidal uptake of radioactive iodine (radioiodide) as an index measurement of thyroidal function. Since the dietary intake of iodide by Americans has increased from approximately 150-200 μg/day in 1950 to 200-750 μg/day after the 1970s, mainly from increased content of iodide in bread, the normal 24-hour thyroidal radioiodine uptake has decreased from 20-50 percent to 10-25 percent. Thyroidal uptake of radioactive iodine is typically increased in the condition known as hyperthyroidism. In this condition, the thyroid gland becomes overactive and produces large quantities of thyroxin, which in turn speeds up all chemical reactions in the body. Radioactive iodine uptake in the thyroid is typically decreased in a condition known as hypothyroidism, but there are some exceptions to this rule. In hypothyroidism, the thyroid produces only small amounts of thyroid hormone, and the body's chemical processes slow down (Greenspan and Rapoport, 1983).

The thyroid secretes ~75 μg of organic (hormonal) iodine daily, mainly as T_4 and a small amount as T_3 or rT_3. Secreted thyroid hormones circulate bound tightly to three blood serum proteins: thyroxine-binding globulin (TBG), thyroxine-binding prealbumin (TBPA, transthyretin) and albumin. Only a small proportion of T_4 (~0.03 percent) and T_4 (~0.3 percent) is free, and it is this small fraction of total thyroid hormones that constitutes the biologically active thyroid hormone (Chopra and Solomon, 1979).

At the present time, blood serum total T_4 concentration is generally measured by a sensitive radioimmunoassay. Radioimmunoassay is the measurement by radioactive detectors of concentrations of radioactive tracer substances in body organs, tissues, or serum. The normal serum T_4 ranges between 5 and 12 micrograms per deciliter (μg/dl). Serum T_4 concentration is increased in hyperthyroidism and decreased in hypothyroidism. It is also increased when serum concentration of TBG, the most avid T_4-binding protein, is increased, and vice versa. Serum TBG is commonly increased in patients taking the hormone estrogen. On the other hand, serum TBG concentration is decreased during treatment with the hormone androgen, or by a congenital defect in TBG synthesis. Like T_4 serum, T_3 concentration can also be measured by a radioimmunoassay. Serum T_3 concentration normally approximates one-seventieth that of T_4 (70-200 ng/dl) in serum. T_3 is increased in hyperthyroidism and is normal or decreased in hypothyroidism (Greenspan and Rapoport, 1983).

In the 1950s, the above-mentioned sensitive and specific measurements of thyroid hormones were not available. Instead, circulating thyroid hormone level was estimated from the measurement of protein-bound iodine (PBI). This measurement took advantage of two characteristics of thyroid hormones: (1) they circulate predominantly (greater than 99 percent) bound to serum proteins; and (2) iodine constitutes the bulk of the weight of thyroid hormone (65 percent in the case of T_4 and 59 percent in the case of T_3). Serum PBI level is increased in hyperthyroidism and decreased in hypothyroidism. Contamination with iodide frequently caused a (misleading) elevation of PBI as a test of thyroid function (Quimby et al., 1958; Blahd, 1965).

Besides the measurement of serum thyroid radioactive iodide uptake (RAIU), radioiodine was used to estimate thyroidal activity by measurement of other parameters in the AAL thyroid study. These parameters included urinary excretion of radioiodine (I^{131}), ratio of salivary excretion of I^{131} to total plasma I^{131} and plasma PBI^{131}, conversion ratio, and plasma levels of PBI^{131}. The tests are discussed below because they are pertinent to the study under examination; a more detailed description of the methodology is available elsewhere (Chopra and Solomon, 1979; Quimby et al., 1958). Such tests are not employed for evaluation of thyroid function today because better methods are now available.

URINARY AND SALIVARY EXCRETION STUDIES

Since iodide circulating in the body is ultimately disposed of by two competing mechanisms, thyroidal uptake and urinary excretion, urinary excretion bears a reciprocal relationship to thyroidal uptake and is an indirect measure of thyroid function. Thus, a higher amount of excreted iodide in the urine is an indication of reduced uptake in the thyroid. Urinary excretion of less than 30 percent of the tracer dose (5 to 50 microcuries [μCi] of I^{131}) in 24 hours suggests hyperthyroidism because the thyroid may be harboring more of the iodide for production of thyroid hormones. Excretions in excess of 40 percent are usually associated with normal or decreased thyroid function. There is a significant degree of overlap in test results between normal subjects and hyperthyroid or hypothyroid patients.

Salivary I^{131} and ratio of salivary I^{131} to total plasma I^{131} and plasma PBI salivary iodide reflect the level of plasma inorganic iodide. When radioiodine is administered to hyperthyroid patients, plasma is rapidly cleared of inorganic iodide and therefore the concentration of radioiodine is decreased in the saliva. In hypothyroidism, radioiodine persists in the plasma and therefore the salivary excretion of radioiodine is increased (Chopra and Solomon, 1979; Quimby et al., 1958).

CONVERSION RATIO AND PLASMA LEVELS OF PBI^{131}

These tests depend on the conversion of iodide trapped by the thyroid to organically bound thyroid hormone, and the secretion of this hormone into the circulation of the bloodstream. There, as noted previously, it is bound by serum proteins and can therefore be separated from iodide by a variety of methods. The conversion ratio is determined at a fixed time, usually 24 hours, after oral administration of a tracer dose ($\sim$50 μCi) of radioiodine. It is expressed as a ratio (in percent) of protein-bound radioiodine in the plasma to the total plasma radioiodine. The ratio is increased in hyperthyroidism and decreased in hypothyroidism. Renal (kidney) insufficiency and congestive heart failure reduce renal clearance of radioiodine and would diminish the diagnostic value of this test in hyperthyroidism.

Plasma levels of PBI^{131} are measured typically at about 72 hours after oral administration of ~25 μCi of radioiodine I^{131}. Levels are expressed as a percentage of the administered dose per liter of plasma. PBI^{131} is elevated in hyperthyroidism and diminished in hypothyroidism.

RACIAL DIFFERENCES IN THYROID FUNCTION

The effect of race has been examined carefully only in a limited number of studies. While in a very few instances variations have been found on the basis of racial differences, other causes may play a role.

In one large study of 1,020 men and 575 women in Jerusalem, serum T_4 was higher in females than in males. Similarly, serum T_4 was higher in 17-year-old boys than in adult men. Serum T_4 was also higher in Israelis originating from Asian and Middle Eastern countries than in those immigrating from Western countries (Slater et al., 1984). In a second study, serum TSH level was found to be lower in American blacks than in whites after adjustment for age and sex, and the difference explained ~6.5 percent of the variation in TSH levels (Schectman et al., 1991). In a third study, prevalence of hypothyroidism was greater in whites than in blacks, in women, and in subjects over 75 years of age (compared with the 55-64-year-old age group (Bagchi, et al., 1990).

Thyroid hormone and/or TSH levels are altered in subjects in some regions because of an effect of prevalent iodine deficiency (Cooper et al., 1993) or abnormalities in serum thyroid hormone-binding proteins (Dick and Watson, 1980). Thus, serum concentration of TBG is much lower in Aborigines in western Australia than in white Australians, and this is associated with decreased serum concentration of total thyroid hormone levels.

The AAL study by Rodahl and his Air Force colleagues (Rodahl and Bang, 1957) did not demonstrate a significant clear difference in thyroidal RAIU, salivary or urinary excretion of radioiodine, or the conversion ratio and PBI^{131} of coastal Inupiats and Athabascan Indians compared with white soldiers and airmen. The RAIU was elevated in inland Inupiats and mountain Athabascan Indians, and this was attributed to decreased intake of iodide in these populations. Rodahl did not observe a significant effect of seasons on thyroid function parameters. More recently, Tkachev and colleagues (1991) have studied the effect of daylight duration on serum TSH and thyroid hormone levels in polar inhabitants. They noted that serum TSH and T_3 increased and serum T_4 and cholesterol decreased with the lengthening of daylight duration. Reed, Brice, and colleagues (1990a) have recently described increased serum binding, yet the increased metabolic clearance rate and daily production rate of T_3 increased in white men who lived in Antarctica for more than five continuous months; the authors named this change the "polar T_3 syndrome" (Reed et al., 1990).

UTILIZATION OF I^{131} AS A DIAGNOSTIC TOOL

I^{130} was the first radioiodine to be employed in clinical studies and has a half-life of only 12.6 hours. It was followed by the use of I^{131}, with a half-life of eight days, which could be produced relatively inexpensively in large quantities; its useful gamma radiation made it the isotope most commonly used in clinical investigation in the 1950s. I^{125} was tested in thyroid studies in the 1960s; its lower radiation energy caused less radiation exposure to the patient than I^{131}, even though I^{125} has a much longer half-life (60 days as opposed to 8 days). However, absorption and retention of I^{125} by overlying tissue was its major disadvantage over the use of I^{131}. I^{131} has now been replaced for clinical use by I^{123}. This radioisotope causes much less radiation exposure to the patient because it has a short half-life (13 hours) and it emits very few or no beta particles during that time.

The dose of I^{131} used for thyroidal radioactive iodide uptake varied with the sensitivity of detecting equipment. Approximately 50-250 μCi was employed when Geiger-Muller tubes were used as detectors. With the availability of efficient scintillation systems with wide-angle amiable radiation collection systems and markedly enhanced sensitivity, it became possible to reduce the dose of I^{131} needed for thyroid uptake to about 5 μCi. The dose of I^{131} recommended for measurement of salivary I^{131} excretion, conversion ratio and PBI^{131} approximated 50 μCi, and this dose was used in the AAL thyroid experiments. I^{131} has also been used for scanning of the thyroid to examine the morphology of the thyroid, and also to evaluate the relative function of different portions of the thyroid. The dose recommended for this purpose approximated 300 μCi (Quimby et al., 1958; Bauer et al., 1953).

CONTRAINDICATIONS TO THE USE OF I^{131}

Pregnancy and lactation have been considered absolute contraindications for the use of I^{131} in treatment of hyperthyroidism (Edmonds and Smith, 1986). No such guidelines are described for diagnostic use of microcurie quantities of I^{131} in pregnant and lactating women. However, most medical centers have avoided use of I^{131} in pregnant and lactating women, and this was also the practice during the period of the 1950s AAL study. Literature review suggests that 400 rads or more of radiation dose is required to cause significant oocyte loss from the gonads (Baker, 1971). This degree of radiation exposure occurs only after therapeutic doses of I^{131}, not at the levels recorded in the AAL experiments. Interestingly, fertility has been observed to remain normal even after therapeutic dosage treatments (Edmonds and Smith, 1986). One case has been described in a woman who was administered I^{131} at therapeutic levels inadvertently during the first week of pregnancy and a high concentration of radioactivity was observed in the pregnant uterus (Cox et al., 1990). Exposure to radiation was considered to be a factor involved in the occurrence of abortion, in this case, at eight weeks of gestation (Romney et al., 1989). Radioiodide is transported freely across the placenta into the fetus. Administration of millicurie doses of I^{131} during pregnancy can adversely affect and even ablate the fetal thyroid after 10 weeks of gestation, when the fetal thyroid becomes capable of accumulating iodine (Green et al., 1971; Shepard, 1967). In men, a high dose of I^{131} ($\sim$350 mCi), as needed for treatment of thyroid cancer, was reported to have been followed by occurrence of testicular failure in one case (Ahmed and Shalet, 1985).

Radioiodide is concentrated in the breast tissue, especially during lactation (Bauemler, 1986; Dydeck and Blue, 1988; Hedrick et al., 1987; Lawes, 1992; Romney et al.; 1989), and therefore it has been recommended that a mother stop breastfeeding up to 14 days after radioiodide has been administered to her (Lawes, 1992; Romney et al., 1989). Some suggest that no radioiodine studies should be done in women who wish to continue breastfeeding (Dydeck and Blue, 1988).

Appendix B
Summary of the Public Session

As the Air Force's Arctic Aeromedical Laboratory (AAL) thyroid function study happened 40 years ago, reconstructing the events of the research and locating the researchers and subjects was a formidable task. The Committee's effort consisted of a detailed historical records search; requests for assistance from local, tribal, state, and federal agencies and governments; telephone and personal interviews; written interviews using a questionnaire; and a public information-gathering session. The purpose of the public session was to obtain information on (1) medical practice and health concerns in Alaska in the 1950s and the 1990s, (2) the role and day-to-day function of the AAL in the 1950s time period, (3) further elaboration on the methodology and conduct of the medical experiments, and, (4) first-hand accounts of the understanding of the participants (both researchers and subjects) about the medical, ethical, and bodily risk elements of that research.

The public session occurred over a two-day period on July 7 and 8, 1994. Several witnesses were identified in advance and requested to provide information during the session. Because of time and distance constraints, some witnesses unable to travel to Fairbanks provided their testimony by telephone conference during the session. Interested members of the public were also invited to provide information.

Subjects of the experiments were asked a standard list of questions about how the study was done, how they were picked to be in the study, what they were told by the Air Force doctors before, during, and after the study, if the doctors spoke with village elders or the witnesses (or their parents if they were very young) to get consent to do the study and what they said to them, if they were told there were any possible ill effects that might be suffered from participating in the study, and if the experiment physicians or other doctors from the Air Force or the Alaska Native Service (ANS) saw the subjects after the study was completed or gave any medical care as a result of the subjects' taking part in their study. These questions were included in a letter sent to each potential witness and an announcement sent to the Inupiat villages.

Staff and researchers from the AAL were asked about the working conditions and organization of the AAL during the period 1955-1957. They were also asked to speak about the state of medical practice in local communities, the relationship between the ANS and the staff of the AAL, methods used to get military and Native volunteers for medical studies, what guidance on informed consent may have existed in the AAL at that time, and what they could remember about studies conducted by Drs. Rodahl and Bang. This testimony was requested in order to provide necessary background primarily on the role and day-to-day function of the AAL in the 1950s time period.

The following discussion summarizes, but does not include in complete detail, the information obtained by guests at the public session.

Information was provided by the Director of the Alaska Division of Public Health of the Department of Health and Social Services on State of Alaska current health concerns and programs, and its scientific and ethical concerns regarding human medical experimentation of Alaskans and military men during the period of the Cold War, and the iodine I^{131} studies in particular. He provided information to the Committee on the efforts of the state to improve health conditions in Alaska. He also provided detailed information on the changes in causes of death between the 1950s and the 1990s, but was unable to provide a breakout of information on thyroid cancers and related deaths in the state.

A retired Director of the Public Health Service Alaska Native Medical Center in Anchorage who is currently a professor at the University of Alaska in Anchorage spoke to the Committee on the subject of 1950s medical practice in Native and local communities of northern and central Alaska. He provided extended testimony on the functioning of the AAL and the ANS, as well as the existence of private and public medical care in the state for the Native communities during that time. His own expertise included knowledge of the history of medical practice in Inupiat populations; he is the compiler of an extensive bibliography and several articles on the subject.

Two former researchers from the AAL now living in Fairbanks and a researcher now in Los Angeles provided testimony to the Committee about the working conditions and organization of the AAL during the period 1955-1957. They told the Committee about the relationship between the ANS and the staff of the AAL, methods used to get military and Native volunteers for medical studies, what guidance on informed consent may have existed in the AAL at that time, the organization and running of the analytical laboratories at the AAL, and what they remembered about studies conducted by Dr. Rodahl. One of the speakers was the director of the analytical laboratory at the AAL in the 1950s, another worked as a technician in the physiology laboratory with Dr. Rodahl at the time of the I^{131} experiments, and the third had conducted extramural research at the AAL during the late 1950s. The former technician recalled that the study had originated from the observation that some individuals of the Alaskan interior showed evidence of modest thyroid enlargement and Dr. Rodahl decided to study it to learn about its causes.

A representative of the Mayor's Office of Point Hope village told of the frustration in his community because there was no information on who had participated in the I^{131} study from Point Hope. He explained that his village had a high rate of cancers, birth defects, and illness and did not know why. Point Hope, he said, was the site of the Project Chariot[1] radiation study in the 1960s, and the villagers were also worried about the effects of radiation from Russian radioactive waste disposal in the ocean where his people fish and hunt.

A physician from the Chief Andrew Isaacs Health Center of the Tanana Chiefs provided extensive information on the health services in the interior of Alaska, the difficulties of modern health care delivery in such a rural area, and the most significant current causes of illness and death. She stated that there had been 11 cases of thyroid cancer since 1969 in the hospital records and two deaths from that disease; she had no way of finding the incidence of goiter for the area her health center serviced, but said it was a common problem. With respect to the matter of informed consent, she said that while there are signed consents required now for medical procedures, communication is still a big problem as of 1994—there are no words in the Athabaskan language for radiation or for cancer, and many patients do not understand what they are told by medical specialists. The English language was very late coming into the region, she said.

A brother and sister who were 20 year old and 17-year-old (respectively) subjects of the AAL I^{131} medical studies in Arctic Village spoke to the Committee about the conduct of the studies, including the potassium iodide control experiment which they both participated in, and a previous AAL study they took part in during 1952. They also spoke to the issue of not knowing how the subjects in Arctic Village were selected for the study, the inability of their parents (their father was the village leader) and themselves to speak English at the time and thus to understand what they were agreeing to, their current physical ailments (including thyroid disease) and the illnesses of their children and grandchildren, and the lack of medical follow-up when the AAL research was completed. They testified they did not know what they were being asked to do by the doctors, but if they understood what was happening, they would not have agreed to participate. Both said that they had respect for and trusted the physicians; they thought the doctors had come to help them back in the 1950s, but now they said it was not true.

A male subject who was originally from Fort Yukon spoke to the Committee about his experience as a subject in medical studies. In great detail he explained how, as an employee of the only hospital in his area in Alaska in the 1950s, he was selected as a subject for three separate AAL studies, including a study which required the surgical implantation and removal of sponges.[2] He said he also participated in a fourth AAL study and an Atomic Energy

---

[1] Project Chariot was an environmental study carried out by the U.S. Geological Survey in 1962 at the behest of the Atomic Energy Commission to examine the movement of radioactive materials through tundra. The study was conducted at Point Hope, Alaska, which was a proposed site for the dredging of a harbor using atomic bombs under the "Atoms for Peace" program. Radioactive fallout from the Nevada test site was placed on the ground in Alaska, watered, and traced for a period of days; the resulting contaminated soil was dug up and put in a mound, then marked radioactive. Local opposition to the site's presence resulted in it being removed in 1994. The soil was shipped to Hanford, Washington, for disposal by the Department of Energy.

[2] The historical report of the AAL for 1 July 1956 to 31 December 1956 (U.S. Air Force, 1956b, p. 37) included reference to the removal of sponges implanted four months previously in subjects from Fort Yukon. The sponges were collected for a study on arteriosclerosis.

Commission (AEC) study in the early 1960s. He said he was not told the capsule given to him in the I^{131} study was radioactive, even though he understood what radioactivity was at the time, having read a book about Marie Curie many years before, and spoke English; he said he was given the opportunity to decline, but was paid to participate in the I^{131} study and wanted to take the test.

The daughter of a woman participant from Arctic Village testified next. She said that her mother, who was too ill to speak before the Committee, had uterine cancer in the early 1950s and was chronically ill, yet was chosen for the I^{131} study. The mother never spoke English but she had told her daughter she thought the AAL physicians had come to help the villagers, and had trusted them. The daughter mentioned the villager's concerns about their illnesses and said that in 1964 the AEC tested the community for exposure to radiation from atomic testing, particularly from eating caribou which had eaten irradiated lichens; she related how one woman who had the highest radiation accumulation had died at an early age. On questioning by the Committee, she said that one woman in her village may have been pregnant at the time she was given the I^{131}. She also expressed outrage at the way Alaska Natives had been studied "like guinea pigs."

Three concerned citizens knowledgeable in the recent history of Arctic Village and/or this particular study provided testimony to the Committee on the Alaska Natives' participation and difficult living conditions. One of the witnesses supported the contention that one woman from Arctic Village was pregnant at the time of the AAL study, and then mentioned that the mother of the brother and sister who had testified earlier in the day had been blind when the AAL tests were conducted on her.

Written testimony was provided to the Committee by the Mayor of the North Slope Borough and Senator Murkowski of Alaska. Both statements were read to the Committee at the public sessions. The Senator, in a letter mentioned in the Preface, shared his concerns about the AAL study and what he would like to see come out of the National Research Council/Institute of Medicine (NRC/IOM) study. The mayor told of his people's worries regarding the different radiation problems his people were exposed to: the I^{131} study, Project Chariot in Point Hope, and Russian dumping of nuclear waste in the Arctic Ocean. The mayor also asked that the entire Committee travel to the other affected villages and talk to all of the I^{131} subjects, plus he requested to review the draft of this report.[3] He stated the need to learn why his people were dying of cancers and other sicknesses and not knowing what the causes were.

Dr. Kaare Rodahl spoke to the Committee by phone from Norway on the conduct of the AAL study, his training in use of I^{131}, training of other physicians in the use of radioisotope administration, the methods used in obtaining Native and military participants' consent, the oversight responsibility of his commanding officer and the Surgeon General of the Air Force,

---

[3]Panel member availability, timing of the study, and funding prevented the full Committee from travelling to each village. However, the Committee and NRC/IOM staff did make a good-faith effort to contact and obtain information from each known living subject it could locate. The draft report could not be made available to the North Slope Borough as this is contrary to NRC/IOM report policy. However, the draft report received rigorous peer review, including reviewers familiar with Native concerns in northern Alaska.

his relationship with the ANS, the issue of informed consent, and how he viewed his working relationship with the Native communities. Of particular note to the Committee was his statement that he did not consider the I^{131} to be radiation. In his words, it was a medical tracer and was given in such small amounts that he did not need to explain it to the research subjects as a radioactive substance. He also said he admired the Native peoples for their ability to live in such harsh and cold conditions, which was the reason he wanted to see if they had a special adaptation to live in their environment. He told us that from 1950 to 1957 (except for 1952 to 1954, when he returned to Norway) he conducted, directed, and oversaw several of studies on acclimatization to cold. The studies, he said, examined the nutrition, physiology, and living habits of Alaska Natives from a variety of villages. He also said that based on conclusions from previous studies by the AAL that found elevated basal metabolism in the Alaska Natives, he proposed to determine the role of the thyroid in man during cold exposure. In his conversation with the Committee, he also was quite clear in stating that two questions he wanted to answer were if the cold weather somehow stimulated the Natives' thyroid glands and if their thyroid activity was high because of endemic goiter.

In the telephone interview at our public meeting, Dr. Rodahl told the Committee that he received the permission of his commanding officer at the AAL to conduct the study, and the headquarters in Washington was also aware of what he was planning to do. He told us that in order to obtain volunteers for his research he approached the commanding officer (in the case of the military participants) or the village elders (in the case of the Alaska Natives) and explained what was to be done, then asked them to bring the volunteers to him. He said he explained what the purpose of his work was and told them they had the right to refuse to participate, using the village elder as an interpreter with the Native subjects. When he was asked about how he described the radioactive tracer, he told us he did not say it was radioactive because in his opinion it was a harmless medical substance that posed no risk to the subjects due to the small amount of radiation involved.

Dr. Rodahl stated in his conversation with us that his research found that salt supplies in the interior villages were not iodized, and a benefit of his work was that it called attention to the problem and the salt supply in the local stores was replaced with iodized salt. He mentioned also that he was in direct contact with the physicians of the ANS who were aware of a goiter problem in those communities, as thyroidectomies had been carried out on some interior Alaskan village members. He told us that in designing his iodine supplement experiment, he was trying to find out if the cold weather was affecting their thyroid activity or if there was some other cause for elevated thyroid activity—in this regard he tested some subjects who had enlarged thyroids on purpose. Dr. Rodahl said when his plane landed, the villagers would come running to meet him and the other physicians who came with him, and the villagers would immediately want their ailments treated. He said the physicians treated them because they were "medical men." He also said the Natives trusted them and they trusted the Natives. However, he also said the AAL physicians were researchers and believed their study would be of benefit to the Natives by helping understand how the thyroid might be stimulated by cold and what might be the cause of goiter in some of the communities.

One former serviceman was contacted by telephone and asked to provide information on his military tour of duty in Fairbanks and his participation in the AAL experiments. He was unable to remember details of taking the I^{131}, but he did recall the method for volunteering, consent for participation, and the physical conditions of the exposure to cold.

The session was open to the public, and questions were allowed of the participants, including those speaking by telephone, by Committee members and members of the audience. Often, the Committee members were questioned by the audience on various technical subjects regarding radiation, radioisotopes, and medical research in order to obtain better knowledge of the I^{131} study. In many cases, the session became an information exchange for the lay audience to express their concerns over radiation in the Arctic in general, or the actions of the government.

Appendix C

Thyroid Radiation Dose Estimates for I^{131} as Determined by the Radiation Internal Dose Information Center, Oak Ridge Institute for Science and Education

Population	Estimated Thyroid Dose[1] (rad/μCi)
Anaktuvuk Pass	2.9
Arctic Village	3.8
Fort Yukon	1.3
White military	1.3
Other[2]	1.3

[1]Radiation dose to the thyroid gland (rad) per unit radioactivity of I^{131} administered (microcurie [μCi]). Estimates are based on data provided in Arctic Aeromedical Laboratory Technical Report 57-36 and the compartmental thyroid model of Stabin et al. These estimates have been used in Table 2.3 to calculate average thyroid dose and thyroid cancer risk.

[2]Thyroid doses could not be estimated for Wainwright, Point Lay, or Point Hope subjects because of insufficient data in Arctic Aeromedical Laboratory Technical Report 57-36. Estimates for these subjects were based on measurements in the Cristy & Eckerman phantom (Report ORNL/TM-8381/VI and V7) and data from the Medical Internal Radiation Dose Committee Dose Estimate Report No. 5 as published in *Journal of Nuclear Medicine* (1975; 16:857-60), assuming 25 percent thyroid uptake of administered I^{131}.

Appendix D

SECRETARY OF DEFENSE[1]
Washington

26 Feb 1953

MEMORANDUM FOR THE SECRETARY OF THE ARMY
SECRETARY OF THE NAVY
SECRETARY OF THE AIR FORCE

SUBJECT: Use of Human Volunteers in Experimental Research

1. Based upon a recommendation of the Armed Forces Medical Policy Council that human subjects be employed, under recognized safeguards, as the only feasible means for realistic evaluation and/or development of effective preventive measures of defense against atomic, biological or chemical agents, the policy set forth below will govern the use of human volunteers by the Department of Defense in experimental research in the fields of atomic, biological and/or chemical warfare.

2. By reason of the basic medical responsibility in connection with the development of defense of all types against atomic, biological and/or chemical warfare agents, Armed Services personnel and/or civilians on duty at installations engaged in such research shall be permitted to actively participate in all phases of the program, such participation shall be subject to the following conditions:

a. The voluntary consent of the human subject is absolutely essential.

(1) This means that the person involved should have legal capacity to give consent; should be so situated as to be able to exercise free power of choice, without the intervention of any element of force, fraud, deceit, duress, over-reaching, or other ulterior form of constraint or coercion; and should have

[1] Retyped for this report from illegible copy.

sufficient knowledge and comprehension of the elements of the subject matter involved as to enable him to make an understanding and enlightened decision. This latter element requires that before the acceptance of an affirmative decision by the experimental subject there should be made known to him the nature, duration, and purpose of the experiment; the method and means by which it is to be conducted; all inconveniences and hazards reasonably to be expected; and the effects upon his health or person which may possibly come from his participation in the experiment.

(2) The consent of the human subject shall be in writing, his signature shall be affixed to a written instrument setting forth substantially the aforementioned requirements and shall be signed in the presence of at least one witness who shall attest to such signature in writing.

(a) In experiments where personnel from more than one Service are involved the Secretary of the Service which is exercising primary responsibility for conducting the experiment is designated to prepare such an instrument and coordinate it for use by all the Services having human volunteers involved in the experiment.

(3) The duty and responsibility for ascertaining the quality of the consent rests upon each individual who initiates, directs or engages in the experiment. It is a personal duty and responsibility which may not be delegated to another with impunity.

b. The experiment should be such as to yield fruitful results for the good of society, unprocurable by other methods or means of study, and not random and unnecessary in nature.

c. The number of volunteers used shall be kept at a minimum consistent with item b, above.

d. The experiment should be so designed and based on the results of animal experimentation and a knowledge of the natural history of the disease or other problem under study that the anticipated results will justify the performance of the experiments.

e. The experiment should be so conducted as to avoid all unnecessary physical and mental suffering and injury.

Downgraded to UNCLASSIFIED 22 August '75 per S. Clements

f. No experiment should be so conducted where there is an a priori reason to believe that death or disabling injury will occur.

g. The degree of risk to be taken should never exceed that determined by the humanitarian importance of the problem to be solved by the experiment.

h. Proper preparation should be made and adequate facilities provided to protect the experimental subject against even remote possibilities of injury, disability, or death.

i. The experiment should be conducted only by scientifically qualified persons. The highest degree of skill and care should be required through all stages of the experiment of those who conduct or engage in the experiment.

j. During the course of the experiment the human subject should be at liberty to bring the experiment to an end if he has reached the physical or mental state where continuation of the experiment seems to him to be impossible.

k. During the course of the experiment the scientist in charge must be prepared to terminate the experiment at any stage, if he has probable cause to believe, in the exercise of the good faith, superior skill and careful judgment required of him that a continuation of the experiment is likely to result in injury, disability, or death to the experimental subject.

l. The established policy, which prohibits the use of prisoners of war in human experimentation, is continued and they will not be used under any circumstances.

3. The Secretaries of the Army, Navy and Air Force are authorized to conduct experiments in connection with the development of defenses of all types against atomic, biological and/or chemical warfare agents involving the use of human subjects within the limits prescribed above.

4. In each instance in which an experiment is proposed pursuant to this memorandum, the nature and purpose of the proposed experiment and the name of the person who will be in charge of such experiment shall be submitted for approval to the Secretary of the military department in which the proposed experiment is to be conducted. No such experiment

Downgraded to UNCLASSIFIED 22 August '75 per S. Clements

shall be undertaken until such Secretary has approved in writing the experiment proposed, the person who will be in charge of conducting it; as well as informing the Secretary of Defense.

5. The address(e)s will be responsible for insuring compliance with the provisions of this memorandum within their respective Services.

/signed/
C. E. Wilson

Copies furnished:
Joint Chiefs of Staff
Research and Development Board

Downngraded to UNCLASSIFIED 22 August '75 per S. Clements

Appendix E
Informed Consent Elements of Disclosure[1]

§—116 General requirements for informed consent.

Except as provided elsewhere in this policy, no investigator may involve a human being as a subject in research covered by this policy unless the investigator has obtained the legally effective informed consent of the subject or the subject's legally authorized representative. An investigator shall seek such consent only under circumstances that provide the prospective subject or the representative sufficient opportunity to consider whether or not to participate and that will minimize the possibility of coercion or undue influence. The information that is given to the subject or the representative shall be in language understandable to the subject or the representative. No informed consent, whether oral or written, may include any exculpatory language through which the subject or the representative is made to waive or appear to waive any of the subject's legal rights, or releases or appears to release the investigator, the sponsor, the institution or its agents from liability for negligence.

(a) Basic elements of informed consent. Except as provided in paragraph (c) or (d) of this section, in seeking informed consent the following information shall be provided to each subject:

(1) A statement that the study involves research, an explanation of the purposes of the research and the expected duration of the subject's participation, a description of the procedures to be followed, and identification of any procedures which are experimental;

(2) A description of any reasonably foreseeable risks or discomforts to the subject;

(3) A description of any benefits to the subject or to others which may reasonably be expected from the research;

(4) A disclosure of appropriate alternative procedures or courses of treatment, if any, that might be advantageous to the subject;

[1]Federal Policy for the Protection of Human Subjects (Federal "Common Rule"), Federal Register, Vol. 56, No. 116 Tuesday, June 18, 1991:pp. 28016-17 (Re-typed)

(5) A statement describing the extent, if any, to which confidentiality of records identifying the subject will be maintained;

(6) For research involving more than minimal risk, an explanation as to whether any compensation and an explanation as to whether any medical treatments are available if injury occurs and, if so, what they consist of, or where further information may be obtained;

(7) An explanation of whom to contact for answers to pertinent questions about the research and research subjects' rights, and whom to contact in the event of a research-related injury to the subject; and

(8) A statement that participation is voluntary, refusal to participate will involve no penalty or loss of benefits to which the subject is otherwise entitled, and the subject may discontinue participation at any time without penalty or loss of benefits to which the subject is otherwise entitled.

(b) Additional elements of informed consent. When appropriate, one or more of the following elements of information shall also be provided to each subject:

(1) A statement that the particular treatment or procedure may involve risks to the subject (or to the embryo or fetus, if the subject is or may become pregnant which are currently unforeseeable;

(2) Anticipated circumstances under which the subject's participation may be terminated by the investigator without regard to the subject's consent;

(3) Any additional costs to the subject that may result from participation in the research;

(4) The consequences of a subject's decision to withdraw from the research and procedures for orderly termination of participation by the subject;

(5) A statement that significant new findings developed during the course of the research which may relate to the subject's willingness to continue participation will be provided to the subject; and

(6) The approximate number of subjects involved in the study.

(c) An IRB may approve a consent procedure which does not include, or which alters, some or all of the elements of informed consent set forth above, or waive the requirement to obtain informed consent provided the IRB finds and documents that:

(1) The research or demonstration project is to be conducted by or subject to the approval of state or local government officials and is designed to study, evaluate, or otherwise examine: (i) Public benefit of service programs; (ii) procedures for obtaining benefits or services under those programs; (iii) possible changes in or alternatives to those programs or procedures; or (iv) possible changes in methods or levels of payment for benefits or services under those programs; and

(2) The research could not practicably be carried out without the waiver or alteration.

(d) An IRB may approve a consent procedure which does not include, or which alters, some or all of the elements of informed consent set forth in this section, or waive the requirements to obtain informed consent provided the IRB finds and documents that:

(1) The research involves no more than minimal risk to the subjects.

(2) The waiver or alteration will not adversely affect the rights and welfare of the subjects.

(3) The research could not practicably be carried out without the waiver or alteration; and

(4) Whenever appropriate, the subjects will be provided with additional pertinent information after participation.

(e) The informed consent requirements in this policy are not intended to preempt any applicable federal, state, or local laws which require additional information to be disclosed in order for informed consent to be legally effective.

Appendix F
Principles for the Conduct of Research in the Arctic[1]

INTRODUCTION

All researchers working in the North have an ethical responsibility toward the people of the North, their cultures, and the environment. The following principles have been formulated to provide guidance for researchers in the physical, biological, behavioral, health, economic, political, and social sciences and in the humanities. These principles are to be observed when carrying out or sponsoring research in Arctic and northern regions or when applying the results of this research.

This statement addresses the need to promote mutual respect and communication between scientists and northern residents and will contribute to the development of northern science through traditional knowledge and experience.

These "Principles for the Conduct of Research in the Arctic" were prepared by the Interagency Social Science Task Force in response to a recommendation by the Polar Research Board of the National Academy of Sciences and at the direction of the Interagency Arctic Research Policy Committee. This statement is not intended to replace other existing federal, state, or professional guidelines, but rather to emphasize their relevance for the whole scientific community. Examples of similar guidelines used by professional organizations and agencies in the United States and in other countries are listed in the publications.

These principles are to be observed when carrying out or sponsoring research in Arctic and northern regions or when applying the results of this research.

[1]Reprinted from *Arctic Research of the United States*, vol. 6, fall 1992.

IMPLEMENTATION

All scientific investigations in the Arctic should be assessed in terms of potential human impact and interest. Social science research, particularly studies of human subjects, requires special consideration, as do studies of resources of economic, cultural, and social value to Native people. In all instances, it is the responsibility of the principal investigator on each project to implement the following recommendations.

1. The researcher should inform appropriate community authorities of planned research on lands, waters, or territories used or occupied by them. Research directly involving northern people or communities should not proceed without their clear and informed consent. When informing the community and/or obtaining informed consent, the researcher should identify:

a. All sponsors and sources of financial support;

b. The person in charge and all investigators involved in the research, as well as any anticipated need for consultants, guides, or interpreters;

c. The purposes, goals, and time frame of the research;

d. Data-gathering techniques (tape and video recordings, photographs, physiological measurements, and so on) and the uses to which they will be put; and

e. Foreseeable positive and negative implications and impacts of the research.

2. The duty of researchers to inform communities continues after approval has been obtained. Ongoing projects should be explained in terms understandable to the local community.

3. Researchers should consult with and, where applicable, include northern communities in project planning and implementation. Reasonable opportunities should be provided for the communities to express their interests and to participate in the research.

4. Research results should be explained in nontechnical terms and, where feasible, should be communicated by means of study materials that can be used by local teachers or displays that can be shown in local community centers and museums.

5. Copies of research reports, data descriptions, and other relevant materials should be provided to the local community. Special efforts must be made to communicate results that are responsible to local concerns.

6. Subject to the requirements for anonymity, publications should always refer to the informed consent of participants and give credit to those contributing to the research project.

7. The researcher must respect local cultural traditions, languages, and values. The researcher should, where practicable, incorporate the following elements in the research design:

a. Use of local and traditional knowledge and experience.

b. Use of the languages of the local people.

c. Translation of research results, particularly those of local concern, into the languages of the people affected by the research.

8. When possible, research projects should anticipate and provide meaningful experience and training for young people.

9. In cases where individuals or groups provide information of a confidential nature, their anonymity must be guaranteed both in the original use of data and in its deposition for future use.

10. Research on humans should be undertaken only in a manner that respects their

privacy and dignity:

a. Research subjects must remain anonymous unless they have agreed to be identified. If anonymity cannot be guaranteed, the subjects must be informed of the possible consequences of becoming involved in the research.

b. In cases where individuals or groups provide information of a confidential or personal nature, this confidentiality must be guaranteed both in the original use of data and in its deposition for future use.

c. The rights of children must be respected. All research involving children must be fully justified in terms of goals and objectives and never undertaken without the consent of the children and their parents or legal guardians.

d. Participation of subjects, including the use of photography in research, should always be based on informed consent.

e. The use and disposition of human tissue samples should always be based on the informed consent of the subjects or next of kin.

11. The researcher is accountable for all project decisions that affect the community, including decisions made by subordinates.

12. All relevant federal, state, and local regulations and policies pertaining to cultural, environmental, and health protection must be strictly observed.

13. Sacred sites, cultural materials, and cultural property cannot be disturbed or removed without community and/or individual consent and in accordance with federal and state laws and regulations.

In implementing these principles, researchers may find additional guidance in the publications listed below. In addition, a number of Alaska Native and municipal organizations can be contacted for general information, obtaining informed consent, and matters relating to research proposals and coordination with Native and local interests. A separate list is available from National Science Foundation's Division of Polar Programs.

PUBLICATIONS

Arctic Social Science: An Agenda for Action. National Academy of Sciences, Washington, DC, 1989.

Draft Principles for an Arctic Policy. Inuit Circumpolar Conference, Kotzebue, 1986.

Ethics. Social Sciences and Humanities Research Council of Canada, Ottawa, 1977.

Nordic Statement of Principles and Priorities in Arctic Research. Center for Arctic Cultural Research, Umea, Sweden, 1989.

Policy on Research Ethics. Alaska Department of Fish and Game, Juneau, 1984.

Principles of Professional Responsibility. Council of the American Anthropological Association, Washington, DC, 1971, rev. 1989.

The Ethical Principles for the Conduct of Research in the North. The Canadian Universities for Northern Studies, Ottawa, 1982.

The National Arctic Health Science Policy. American Public Health Association, Washington, DC, 1984.

Protocol for Centers for Disease Control/Indian Health Service Serum Bank. Prepared by

Arctic Investigations Program (CDC) and Alaska Area Native Health Service, 1990. (Available through Alaska Area Native Health Service, 255 Gambell Street, Anchorage, AK 99501.)

Indian Health Manual. Indian Health Service, U.S. Public Health Service, Rockville, Maryland, 1987.

Human Experimentation. Code of Ethics of the World Medical Association (Declaration of Helsinki). Published in British Medical Journal, 2:177, 1964.

Protection of Human Subjects. Code of Federal Regulations 45 CFR 46, 1974, rev. 1983.

Appendix G
Biographical Sketches of Committee Members

Chester M. Pierce earned his AB from Harvard College and his MD from Harvard Medical School and has received honorary Doctorates of Science from Westfield College and Tufts University. He is a professor of psychiatry at Harvard Medical School, Boston, Massachusetts. He was the Chairman of the ad hoc Committee on Polar Biomedicine of the National Research Council's Polar Research Board in the 1980s, andhas been the U.S. delegate for the National Academy of Sciences to the international Scientific Committee on Antarctic Research's Working Group on Human Biology and Medical Science since 1977. Dr. Pierce is Past President of the American Board of Psychiatry and Neurology. He is a member of the Institute of Medicine of the National Academy of Sciences and is a recipient of the Special Recognition Award from the National Medical Association as well as having been made an Honorary Fellow of the Royal Australian and New Zealand College of Psychiatrists. He also has a geographic feature in Antarctica named for him—Pierce Peak.

David R. Baines earned his BS from Arizona State University and his MD from the Mayo Medical School. He is a Board Certified family practitioner in private practice at St. Maries Family Medicine Clinic in St. Maries, Idaho. He has been President of the Executive Board of the Association of American Indian Physicians, he is the current Chair of the Ad Hoc Committee on Minority Populations at the National Heart, Lung and Blood Institute of the National Institutes of Health, and is a recipient of the 1993 Gentle Giant of Medicine Award from G.D. Searle & Company. He was selected for the 1995 United States Public Health Service Primary Care Fellowship and was also selected to be on the Graduate Training and Family Medicine Review Committee in the Division of Medicine, Bureau of Health Professions of the Health Resources and Services Administration. He is on the Commission on Membership and Member Services of the American Academy of Family Physicians and is a member of the National Research Council, Polar Research Board, and is a consultant to the Centers for Disease Control and Prevention and the Department of Health and Human Services. He is a member of the Tlingit and Tsimpshian Tribes of Alaska and works on the Coeur d'Alene Indian Reservation in Idaho. Dr Baines frequently lectures on how he incorporates his culture and traditional beliefs into his practice of modern medicine.

Inder J. Chopra received his MB, BS ('61), and MD ('65) from the All India Institute of Medical Science. He received his American Board diploma in Internal Medicine in 1972 and his certification in endocrinology in 1973. He is a licensed physician in the State of California. He is a Professor of Medicine, School of Medicine at the University of California Los Angeles (UCLA) and a staff physician at the Center for Health Science at UCLA. He is the recipient of the Van Meter-Armour and the Parke-Davis Awards from the American Thyroid Association, the Ernst Oppenheimer Memorial Award from the American Endocrine Society. He is a Fellow of the American College of Physicians, as well as a member of the American Thyroid Association and the Endocrine Society.

Nancy M. P. King is Associate Professor of Social Medicine at the University of North Carolina School of Medicine. She received her BA from St. John's College and her JD from the University of North Carolina School of Law. She is a member of the Maryland Bar and the North Carolina Bar. Professor King teaches legal, social, and ethical issues to medical students as part of a comprehensive social medicine curriculum at UNC. Her research interests center on the study of roles and responsibilities in health care decisions, including issues related to informed consent, neonatal intensive care, the development and use of experimental technologies, and decisionmaking at the end of life. She is a collaborating author of A History and Theory of Informed Consent, and has written about informed consent and decisionmaking in health care treatment and research in the health law and medical ethics literature. A revised edition of her book, Making Sense of Advance Directives, will be published by Georgetown University Press in 1996.

Kenneth L. Mossman received his BS from Wayne State University in Biology, his MS and PhD from the University of Tennessee in Radiation Biology, and MEd from the University of Maryland in Education Policy, Planning, and Administration. From 1990 to 1992 he served as Assistant Vice President for Research at Arizona State University. He is currently Professor, Department of Microbiology, Arizona State University. He is a recipient of the Elda E. Anderson and the Marie Curie Gold Medal Award from the Health Physics Society and is the Health Physics Society's Past President. Dr. Mossman was elected fellow of the Health Physics Society in 1994.

Appendix H
Glossary

basal metabolism	The energy required to keep the body functioning at rest.
beta particles	Charged particles emitted from the atomic nucleus during radioactive decay. Beta particles are identical to electrons. I^{131} emits beta particles.
cholesterol	Chemically a lipid; also an important constituent of body cells; also, among other functions, involved with the formation of hormones.
colloid	A type of liquid similar to a suspension. Thyroid hormones are stored in thyroid gland in the form of colloid.
conversion ratio	A factor used to convert amount of radioactivity administered to absorbed dose in tissue.
dose conversion	See conversion ratio.
dosimetry	Measurement of absorbed dose. A way to monitor amount of radiation exposure.
E-6	Shorthand for $10^{-6} = 1/1{,}000{,}000$ (one in a million)
estrogen	A group of hormones essential for normal female sexual development and for the healthy functioning of the female reproductive system.
excess risk	The risk over and above the natural risk that may be attributable to exposure to some agent.
excretion	The discharge of waste material from the body.
extracellular fluid	Fluid that exists outside a cell.

follicular cells	Cells of thyroid glands engaged in production of thyroid hormone.
gamma radiation	Very-short-wavelength, high-energy electromagnetic radiation emitted by radioactive material during radioactive decay.
Geiger-Muller tube	A device used to measure beta particles and gamma rays.
half-life	A property of radioactive atoms. The half-life is a measure of the rate of decay and is the time necessary for a quantity of radioactive atoms to decay to one-half of the original number. I^{131} has a half-life of eight days.
hyperthyroidism	A condition caused by excessive secretion of the thyroid hormones, which increases the basal metabolic rate and causes an increased demand for nutrients and oxygen to support this metabolic activity.
hypothyroidism	A condition due to deficiency of thyroid secretion. (This condition results in lowered basal metabolism, i.e., obesity.)
I^{131}	Radioactive isotope of iodine with an atomic weight of 131.
I^{130}	Radioactive isotope of iodine with an atomic weight of 130.
I^{125}	Radioactive isotope of iodine with an atomic weight of 125.
I^{123}	Radioactive isotope of iodine with an atomic weight of 123.
initiated cell	An activated cell.
isotope	Forms of an element with the same number of protons in the nucleus and thus the same chemical properties. I^{130}, I^{131}, and I^{123} are isotopes of iodine.
iodide	A compound of iodine with another element, e.g., potassium iodide or sodium iodide.
microcurie	Amount of radioactivity; 1 curie is equal to 37 billion disintegrations per second (dps). 1 microcurie = 1/1,000,000 of 1 curie = 37,000 dps.
milliliter	Unit of volume. 1 milliliter = 1/1,000 of 1 liter.)

mCi	1 mCi = millicurie = 1/1,000 of 1 curie. (See microcurie.)
observed risk	The risk of incidence or mortality of a disease that has been determined through observation or epidemiological studies.
oocyte	The early or primitive ovum (before it has developed completely).
protein-bound thyroid	Thyroid hormones bound to serum proteins.
radiation	Energy in the form of waves or particles.
rad	Unit of absorbed dose. (See cGy. 1 rad = 1 cGy.)
radioimmunoassay	A laboratory technique that employs a radioactive isotope to measure the concentration of specific substances in blood.
radionuclides	A nuclide (a species of atom with a given number of neutrons and protons in its nucleus) that is radioactive.
radiopharmaceuticals	Drugs or medicines that contain a radionuclide.
renal insufficiency	The reduction in the ability of the kidneys to filter waste products from the blood and excrete them in the urine, to control the body's water and salt balance, and to regulate the blood pressure.
retention rate	The rate at which a substance is maintained in living tissue before being removed by excretion or other processes.
reverse T_3	An amino acid, an inactive metabolite of T_4.
roentgen	Unit of radiation exposure.
scaler	An electronic device that can be used with a radiation detector to measure radiation.
scintillation counter	A radiation detector used to measure I^{131} in bodily fluids (e.g., blood).
standard capsule	A formulation of I^{131} containing a known amount of I^{131} (usually 50-65 microcuries of I^{131}).
standard error	A measure of statistical variability of a number of measurements about the mean.

Tc-99m	Technium 99, a radioactive substance commonly used in nuclear medicines.
threshold	A dose below which an effect is not observed.
thyroxine	A hormone produced by the thyroid gland.
tracer dose	A quantity of radioactive material too small to cause adverse health effects.
triiodothyronine	One of two forms of the principal hormone secreted by the thyroid gland.
T_3	One of the hormones made in the thyroid. It is also produced in extrathyroidal tissues such as the liver and kidney.
T_4	One of the hormones made in the thyroid.
μCi	Shorthand for microcurie.
uptake	The process of a tissue or organ accumulating a particular compound or substance. Thyroid uptake refers to accumulation of iodine in the thyroid gland.